THE SYSTEMS VIEW OF THE WORLD

A Holistic Vision for Our Time

ADVANCES IN SYSTEMS THEORY, COMPLEXITY, AND THE HUMAN SCIENCES

Alfonso Montuori, Series Editor

The Systems View of the World
    Ervin Laszlo

forthcoming

Evolution: The Grand Synthesis
    Ervin Laszlo

Homeland Earth: A Manifesto for the New Millenium
    Edgar Morin and Anne Brigette Kern

New Paradigms, Culture and Subjectivity
    Dora Fried Schnitman (ed.)

THE SYSTEMS VIEW OF THE WORLD

A Holistic Vision for Our Time

ERVIN LASZLO

Fourth printing 2002

Printed in the United States of America

Library of Congress Cataloging-in-Publication Data

Laszlo, Ervin, 1932-
    The systems view of the world : a holistic vision for our time
Ervin Laszlo
        p. cm. -- (Advances in systems theory, complexity, and
the human sciences)
    Includes bibliographical references and index.
    ISBN 1-57273-053-6 (pbk. : alk. paper)
    1. System theory. 2. Science--Philosophy I. Title.
II. Series.
Q295.L367   1996
003--dc20   96-711
                                                        CIP

Hampton Press, Inc.
23 Broadway
Cresskill, NJ 07626

Contents

Preface

The man who does not possess the concept of physics (not the science of physics proper, but the vital idea of the world which it has created), and the concept afforded by history and by biology, and the scheme of speculative philosophy, is not an educated man. Unless he should happen to be endowed with exceptional qualities, it is extremely unlikely that such a man will be, in the fullest sense, a good doctor, a good judge, or a good technical expert.

Ortega y Gasset
Mission of the University

ONE could paraphrase Mark Twain in saying that everybody talks about the new worldview of the sciences but nobody quite knows what it is. Certainly, not many people do anything about exploring it and making it explicit. Yet understanding this worldview is important. If you want to change the world, or at least make sure that it doesn't head blindly to its own destruction, you have to understand the nature of this world. And if you want to understand it, you have to interpret what you experience and know of it through some intelligible hypothesis. Unless you have privileged access to ultimate reality through intuition or illumination, you must choose an empirical concept for your understanding—one that is based on how human beings interact with the world around them.

There are many such empirical concepts; all of us hold one or more, more often tacitly than consciously. Ordinary common sense provides one variety of empirical concept, and this changes with the intellectual and emotional sensibilities of the person, his social role, and the character of his culture. The vision of artists and religious leaders offers alternative concepts, more removed from the here-and-now of everyday experience. But the single continuously tested and critically constructed empirical world concept comes from the theories of the contemporary empirical sciences. If we want to understand the world around us, whether because we want to change it, preserve it, or just have the satisfaction of knowing it, we could very likely not do better, and could do a lot worse, than to turn to the contemporary sciences for elucidation.

The contemporary sciences, however, present an awesome front of complex mathematics and ten-syllable words to the uninitiated. The meaning of the theories is obscured by the techniques of checking and improving them. Even the working scientist himself is normally content to leave his theories uninterpreted. He is seeking specific quantitative answers to particular problems, generated in the framework of a multitude of previous answers to foregoing problems. Yet, as leading scientists in every field are now clearly recognizing, there is no theory without an underlying worldview which directs the attention of the scientist. There is no experiment without a hypothesis and no science without some expectation as to the nature of its subject matter. The underlying hypotheses guide theory formulation and experimentation, and they are in turn specified by the results of the experiments designed to test the theories. The problem, both for science and for the widest communities of concerned persons, is to explicate this implicit vision and to focus it on the issues that now dominate our mind and will soon decide our future.

A new view of the world is taking shape in the minds of advanced scientific thinkers the world over, and it offers the best hope of understanding and controlling the processes that affect the lives of us all. Let us not delay, then, in doing our best to come to a clear understanding of it.

1

The Atomistic View and the Systems View

UNTIL very recently, contemporary science was shaped by a mode of thinking which placed rigorous detailed knowledge above all other considerations. This mode of thought was based on the implicit belief that the human mind has a limited capacity for storing and processing information. If you know some things very thoroughly, you can not know very many different kinds of things. If you have some acquaintance with many different things, chances are you do not know them thoroughly. But individual people can work in teams, and what one of them knows can be complemented by the knowledge of the others. Hence knowledge can proceed in depth without thereby losing breadth. This is the ideal of specialization, and it has led to the great advances in the sciences and technologies that now affect the lives of every one of us.

There is one difficulty with specialization, however. This is the tendency of patterns of knowledge to create closed bubbles in their own right. Specialists in one field can communicate with one another if they share a specialty, but experience difficulty when their interests do not coincide. It may be an exaggeration to claim, as the story has it, that a geologist specializing in soft rocks finds himself a lonely man in a congress on hard-rock geology, but it is no exaggeration that geologists and biologists have difficulties understanding each other even when their interests are relatively close. And what goes for specialized disciplines goes also for individual workers within the disciplines. Given persons can and often do develop interests within their own specialties which set them apart from the rest—or all but a handful among the rest—and create a kind of specialty bubble around themselves. This is true not only in the sciences but also in the arts and humanities. The literary historian specializing in early Elizabethan theater may not have much in common with a colleague specializing in Restoration drama, and will find himself reduced to conversation about the weather when encountering an expert on contemporary theater.

The unfortunate consequence of such specialty barriers is that knowledge, instead of being pursued in depth and integrated in breadth, is pursued in depth in isolation. Rather than getting a continuous and coherent picture, we are getting fragments—remarkably detailed but isolated patterns. We are drilling holes in the wall of mystery that we call nature and reality on many locations, and we carry out delicate analyses on each of the sites. But it is only now that we are beginning to realize the need for connecting the probes with one another and gaining some coherent insight into what is there.

The ideal of rigorous information, while it atomized our understanding, did give us a healthy respect for tested knowledge. We are no longer willing to put up with theories which stitch together gaps in knowledge with the fabric of pure faith or imagination. But those who believe that such a patchwork approach is the only means of gaining a coherent and integrated world view today are mistaken. We already have the foundations of such a worldview. It is taking shape as the implicit natural philosophy of the present and next development of the contemporary sciences. There is an emerging paradigm—a new way of ordering the information we already have and are likely to get in the foreseeable future. Let us turn now to a consideration of this new way of looking at the world, and the reasons why it is preferable to the atomistic method of compartmentalized specialization.

1. Why Science Shifts Its Sights

The specialist looks at carefully isolated phenomena: he is interested in how one thing affects another. He can compute the effect by looking at things as separate facts connected by some causal or correlative relationship. This makes it possible to research and systematize many detailed processes in the natural world. One can tell how one cell or organ reacts to one particular kind of stimulant, or how one body reacts to one particular kind of force. We can prescribe medicines and build machines based on this knowledge. And we can continue to fill our storehouses of tested information. But there is one thing such knowledge cannot tell us, and that is how a number of different things act together when exposed to a number of different influences at the same time. And almost everything we encounter around us contains a large number of different things and is exposed to a number of different influences.

We ourselves are composed of some five octillion atoms, and our brains, of ten thousand million neurons. A hydrogen atom is composed of a proton and neutron in its nucleus and one electron in its shell, but the number of forces acting within it are so complex that mathematicians need multidimensional spaces to represent them. And atoms more complex than helium (which has two orbital electrons) contain three or more "bodies" in their shells and our mathematics are incapable of solving the three-body problem—that is, handling equations of motion for more than two objects moving under mutual influence. In other words, we are quite incapable of proceeding with the rigorous techniques of specialization for any phenomenon more complex than a helium atom.

When such problems are handled through the method of piecemeal analysis, they are handled through a simplification: it is assumed that the forces or "bodies" calculated are interacting in sequences of interacting pairs. For many purposes this is a workable assumption. It can give useful information on specific interactions and can aid us in developing techniques for manipulating and predicting them. But special analysis does not give us a true mapping of many things, for real things tend to be far more complex than it can handle.

What, then, are the contemporary sciences doing about this? They offer a solution which is another simplification of the real states of affairs, but one that is more adequate to grasping their complex nature. Instead of looking at one thing at a time, and noting its behavior when exposed to one other thing, the sciences now look at a number of different and interacting things and note their behavior *as a whole* under diverse influences. This is similar to what we do in everyday life when we think of players as teams rather than as interacting individual performers. We do the same thing quite as readily in thinking of business enterprises as companies rather than as individual workers and administrators. In fact, we do it with nations as well, and with classes and groups of people within nations. We even speak of international blocs in simplified yet rather adequate terms. We feel that we can safely disregard the unique individuality of the members of such units as long as there are certain types of members in certain proportions and relationships. It does not matter *who* does this job or that—as long as there is *someone* to do it.

The large groups we thus come to know appear to establish their own "personalities." Even if most of their individual members change, the groups' characteristics tend to be preserved. For example, over the years athletic teams exchange their players, with younger ones replacing the veteran performers. Yet the teams usually maintain much of their own characteristics—their tactics and techniques, their fighting spirit, and so on. Even more striking is the continuity in the case of business corporations, where everyone can be replaced, from the president to the office boy, and yet the company can continue to exist with very much the same features it had before. This is true for entire communities and nations—even entities as large and nebulous as cultures. Individuals come and go; they remain. It is not that they are immune to change themselves, but they do not change with the changes in their membership. It is as though they had a life and personality of their own.

Such characteristics of "wholes" are typical for all groups of interacting parts when the parts maintain some basic sets of relationships among themselves. It does not matter for a carbon atom which electron fills which shell—as long as its "permissible bands of energy" are filled with a number of electrons proportionate to the number of neutrons in its nucleus. This is much like the relative indifference of a corporation as to just who its employees are, as long as there are a sufficient number of persons with sufficient qualifications in the proper relationships to each other and to their tools and instruments. Hence such entities exhibit a certain uniqueness of characteristics as wholes. They cannot simply be reduced to the properties of their individual parts.

Of course, it is quite possible that we could fully account for the properties of each whole if we could know the precise characteristics of all the parts and know in addition all existing relationships between them. Then we could reduce the characteristics of the whole to the sum of the characteristics of the parts in interaction. But this involves integrating the data not merely for three bodies, but for three thousand, three million, three billion, or more, depending on the whole we are considering. And since the sciences cannot perform this feat even for a set of three parts, it is quite hopeless to think that they can do it for any of the more complex phenomena in nature and society. Hence, to all intents and purposes, the characteristics of complex wholes remain irreducible to the characteristics of their parts. This is by far the most indicated working assumption, since not only could we not compute the behavior of the whole from the behavior of its parts, but we would have to revise our computations with every change in "personnel."

Yet by assuming, as a matter of expediency, that groups of interacting parts with maintained basic structure have characteristics of their own, the contemporary sciences succeed remarkably well in explaining and even predicting their behavior. The science of economics, for example, could not possibly take into account the behavior of every single consumer and producer in an economy. Yet it can discern general characteristics of the economy which it formalizes into laws, and these yield predictions. And predictions can be tested against experience.

One of the predictions that (unfortunately) often comes true is the number of highway accidents on a national holiday. Yet the individual drivers on the highways change from year to year and so do their individual abilities, routes, and preoccupations. It would be literally impossible to try to predict the number of fatal accidents on a holiday weekend by analyzing the abilities, routes, and preoccupations of each and every driver on the road. Yet by taking the drivers as members of the class of holiday motorists, and considering past patterns of their behavior together with the state and frequency of roads, number of cars in service, and a few other factors, it is possible to come significantly close to predicting the correct number of highway deaths. It is as though the group of holiday motorists had characteristics of its own, which were not reducible to the characteristics of any individual driver on the road. Certainly there is nothing in anybody's driving skill which would indicate the recurrent regularity of fatal accidents on such occasions. Yet the driving skills, routings, and habits of all people, in mutual relation, add up to these characteristics.

More important examples of the same phenomenon are encountered when we analyze ourselves as wholes composed of a multitude of interacting atoms, molecules, cells, tissues, and organs, and our societies as consisting of a multitude of people in some form of intercommunication. In each case there are sets of relationships which are conserved, even though all participants are replaced sooner or later. The cells of our body are renewed over a period of about seven years, while the individual members of a state are renewed over a period of about seventy. Yet the relationships which make me what I am, and this country what it is, remain unchanged—or rather, the change is much slower, and not fully in function of the change of its parts. I get older in seven years, not because particular cells in my body are no longer part of me, though the number of neurons in my brain do diminish in time, but because the relationship of the cells I possess have undergone some subtle changes which we call aging. And the same goes, *mutatis mutandis*, for teams, armies, corporations, nations, and worldwide organizations. Some individuals make more difference to the character of a nation than others, but none are irreplaceable. It is a precarious organization, indeed, that is tied to the character or personality of a single person however great he or she may be. No organization could survive under such conditions for very long.

2. The Rise of the Systems Sciences

The protosciences of antiquity sought to penetrate the complexities of phenomena by insight or revelation. Their theories were imaginative and sometimes inspired, but they could seldom stand the test of confrontation with actual experience. Modern science insisted on such confrontation, and discarded all theories (such as theologies or concepts of soul) which either could not be tested against experience or failed the test. Since only simple interactions could be definitely tested, modern science developed as the science of Galileo and Newton. It could handle relatively simple relationships between forces or bodies, and it presented a world picture of a universe that is reducible to such relationships in all its essential respects. Newtonian science looked upon the physical universe as an exquisitely designed giant mechanism, obeying elegant deterministic laws of motion. Complex sets of events could be

7

understood by this science only when broken down to their elementary interactions. Whatever was clearly known behaved like a reliable mechanism, and the rest was assumed to do likewise (with the possible exception of mind—a phenomenon which Newtonian science could not even begin to comprehend). Thus the world was thought to *be* a mechanism, made up of a large number of uniformly behaving parts.

The beginning of the twentieth century witnessed the breakdown of the mechanistic theory even within physics, the science where it was the most successful. Sets of interacting relationships came to occupy the center of attention, and these were of such staggering complexity—even within a physical entity as elementary as an atom—that the ability of Newtonian mechanics to provide an explanation had to be seriously questioned. Relativity took over in field physics, and the science of quantum theory in microphysics. The progress of investigation in other sciences followed parallel paths. Biology attempted to divest itself of the *ad hoc* dualism of a "life principle" as it appeared in the vitalism of Driesch, Bergson, and others, and tried to achieve a more testable theory of life. But the laws of physics were insufficient to explain the complex interactions which take place in a living organism, and thus new laws had to be postulated—not laws of "life forces," but laws of integrated wholes, acting as such. Just as the science of economics proved to be incapable of explaining the rise of stock prices on the basis of the individual personalities of stockbrokers and public, so the science of biology was unable to explain the self-preservation of the animal organism by recourse to the physical laws governing the behavior of its atoms and molecules. New laws were postulated, which did not contradict physical laws but *complemented* them. They showed what highly complex sets of things, each subject to the basic laws of physics, do when they act together. In view of parallel developments in physics, chemistry, biology, sociology, and economics, many branches of the contemporary sciences became, in Warren Weaver's phrase, "sciences of organized complexity"—that is, *systems* sciences.

Equipped with the concepts and theories provided by the contemporary systems sciences, we can discern strands of organized complexity wherever we look. We ourselves are a complex organized system, and so are our societies and our environment. Nature itself, as it manifests itself on this earth, is a giant "Gaia" system maintaining itself, although eventually all its individual parts get sifted out and replaced, some more quickly than others. Setting our sights even higher in terms of size, we can see that the solar system and the galaxy of which it is a part are also systems, and so is the metagalaxy of which our galaxy is a component.

Some systems endure for a relatively long time—a stable atom, for example, or the biosphere as a whole. Others are more short-lived, such as a May-fly or a picket line. Yet while they exist, regardless of how long, each system has a specific structure made up of certain maintained relationships among its parts, and it manifests irreducible characteristics of its own. If we want to know more about them we have to treat them as systems, that is, as wholes with properties of their own. That way we can find out something about them—how they behave under various conditions, how they evolve or decay, what parts or subsystems have controlling influence within them, and so on. It is quite unfeasible to come to know these things by considering the specific interactions of each of their individual parts; there are too many of them.

Ours is a complex world. But human knowledge is finite and circumscribed. "Nature does not come as clean as you can think it," warned Alfred North Whitehead, and went on to propound an extremely clean and elegant cosmology. Since theories, like window panes, are clear only when they are clean, and the world does not come as cleanly as all that, we must know where we perform a clean-up operation. Scientific theories, while simpler than reality, must nevertheless reflect its essential structure. Science, then, must beware of rejecting the complexity of structure for the sake of the theory's simplicity; that would be to throw out the baby with the bath water.

The specialists concentrate on detail and disregard the wider structure which gives it context. The systems scientists, on the other hand, concentrate on structure on all levels of magnitude and complexity, and fit detail into its general framework. They discern relationships and situations, not atomistic facts and events. By this method they can understand a lot more about a great many more things than the rigorous specialists, although their understanding is more general and approximate. Yet, some knowledge of connected complexity is preferable even to a more detailed knowledge of atomized simplicity, if it is connected complexity with which we are surrounded in nature and of which we ourselves are a part. If this is the case, to have an adequate grasp of reality we must look at things as systems, with properties and structures of their own. Systems of various kinds can then be compared, their relationships within still larger systems defined, and a general context established. If we are to understand what we are, and what we are faced with in the social and the natural world, evolving a general theory of systems is imperative.

"Systems sciences" are springing up everywhere, as contemporary scientists are discovering organized wholes in many realms of investigation. Systems theories are applied in almost all of the natural and social sciences today, and they are coming to the forefront of the human sciences as well. These new sciences adopt a flexible method. The systems method does not restrict the scientist to one set of relationships as his object of investigation; he can switch levels, corresponding to his shifts in research interest. Systems science can look at a cell or an atom as a system, or it can look at the organ, the organism, the family, the community, the nation, the economy, and the ecology as systems, and it can view even the biosphere—the Gaia system—as such. A system in one perspective is a subsystem in another. But the systems method always treats systems as integrated wholes of their subsidiary components and never as a mechanistic aggregate of parts in isolable causal relations.

3. A Contrast of Worldviews

The holistic vision of the systems sciences contrasts with the atomistic and mechanistic worldview of the classical disciplines. For example:

- The worldview of the classical sciences conceptualized nature as a giant machine composed of intricate but replaceable machine-like parts. The new systems

10

sciences look at nature as an organism endowed with irreplaceable elements and an innate but non-deterministic purpose for choice, for flow, for spontaneity.

- The classical worldview was atomistic and individualistic; it viewed objects as separate from their environments and people as separate from each other and from their surroundings. The systems view perceives connections and communications between people, and between people and nature, and emphasizes community and integrity in both the natural and the human world.

- The classical worldview was materialistic, viewing all things as distinct and measurable material entities. The systems view gives a new meaning to the notion of matter, as a configuration of energies that flow and interact, and allows for probabilistic processes, for self-creativity, as well as for unpredictability.

- In its application to everyday affairs, the classical worldview extolled the accumulation of material goods and promoted a power hungry, compete-to-win ethos. The new vision emphasizes the importance of information and hence of education, communication, and human services over the accumulation of material goods and the acquisition of raw power.

- The classical worldview saw growth in the material sphere as the pinnacle of socioeconomic progress and promoted greater and greater use (and indirectly of waste) of energies, raw materials, and other resources. The systems view, looking first of all to the whole formed by social and economic parts, insists on sustainable development through flexibility and accommodation among cooperative and interactive parts.

- The classical worldview was Eurocentric, taking Western industrialized societies as the paradigms of progress and development. The holistic vision takes in the diversity of human cultures and societies and sees all of them as equally valid, ranking them only in regard to sustainability and the satisfaction they provide for their members.

- The classical worldview was also anthropocentric, perceiving human beings as mastering and controlling nature for their own ends. The systems view sees humans as organic parts within a self-maintaining and self-evolving whole that is the context and the precondition of life on this planet.

- When the classical worldview was applied to social science, the dominant notions turned out to be struggle for survival, the profit of the individual, with at best an assumed automatic coincidence of individual and societal good (through Adam Smith's "invisible hand"). When the systemic vision inspires the theories of social science, the values of competition are mitigated by those of cooperation, and the emphasis on individualistic work ethos is tempered with a tolerance of diversity and of experimentation with institutions and practices that foster man-man and man-nature adaptation and harmony.

- When the classical worldview was applied to medical science, the human body appeared to be a machine frequently in need of repair by factual and impersonal interventions and treatments. The problems of the mind were seen to be separable from those of the body and hence to be separately treated. When the systems view is the basis of a diagnosis the body is seen as a system of interacting parts, and body and mind are not separable. It is the health of the whole system that is to be maintained by attention to psychic and interpersonal as much as to physical and physiological factors.

The shift from the classical to the systemic worldview is healthy, and its completion is urgent. Worldviews are constellations of concepts, perceptions, values, and practices that are shared by a community and direct the activities of its members. A given worldview can be shared by a small community, such as a research team, or a large one, such as an entire culture. It can help people comprehend and explain the nature of the world in which they live, as well as their role and identity in that world. If a worldview is coherent and embracing, it can also provide a pathway for carrying people through the succeeding epochs of their lives, from childhood through adolescence to adulthood and into old age. And if it is consciously held, it can provide guidelines for establishing congenial personal relationships and social roles, and fulfilling patterns of work.

In Western society the mythical worldview of antiquity and the doctrinaire worldviews of the Middle Ages have been surrendered, and the resulting vacuum was to be filled by science. The atomistic worldview inspired by Newtonian science promised to fulfill the functions of comprehensive and consciously held worldviews, and Marxists and other stalwart souls believed that a scientific concept would one day eliminate the need for myth and religion altogether. In our day, however, the promise of a worldview derived from the classical tenets of modern science is increasingly questioned. Alienation and anomie are on the rise, and adherence to an atomistic concept offers scant relief. There is an urgent need to go beyond classical science's view of the world, to a more integrated but no less tested and testable view.

We cannot expect to satisfy all the requirements attaching to a worldview in reference to science alone, without also drawing on the insights of religion and the values of humanism, but we can and should recognize that the avant-garde branches of the contemporary sciences are veritable fountainheads for the creation of a non-atomistic and non-mechanistic vision that can fill the need for practical guidance in our times. The new systems view can provide the clues, the metaphors, the orientations, and even the detailed models for solving critical problems on this precious but increasingly crowded and exploited planet.

2

What is a System?

IN the history of European science, atomistic and holistic ways of thinking have alternated. Early scientific thinking was holistic but speculative; the modern scientific temper reacted by being empirical but atomistic. Neither is free from error, the former because it replaces factual inquiry with faith and insight, and the latter because it sacrifices coherence at the altar of facticity. We witness today another shift in ways of thinking: the shift toward rigorous yet holistic theories. This means thinking in terms of facts and events in the context of wholes, forming integrated sets with their own properties and relationships. Looking at the world in terms of such sets of integrated relations is the current and next choice over atomism, mechanism, and uncoordinated specialization.

Systems thinking gives us a holistic perspective for viewing the world around us, and seeing ourselves in the world. It is a way of organizing, or perhaps reorganizing, our knowledge in terms of systems, systemic properties, and inter-system relationships. Before we explore what worldview results from such an organization of current knowledge, we should clarify a basic question. *Just what is a system?* Unless we understand the nature of systems, the entire worldview that results from our taking it as the key concept will be fuzzy at best. This chapter intends to give a brief, but hopefully clear and meaningful, answer to this question.

When we speak of a "system" we often speak of something that exists only in our mind. For example, a "theological system" or a "system of logic" exists solely in the minds of human beings and not in the world that they inhabit. Traditionally, it was such abstract entities that were intended by the notion of system. This has now changed. Many of the systems that surround us are real-world systems: they have a definite standing in reality, independently of our thinking of them. Such is a political system (as contrasted with a system of political principles), an economic system, a social system, and so on. Similarly, in the case of new-fangled kinds of systems, such as a computer system that involves hard- and software and perhaps a network, and not just the design for them.

Even more remarkably, we now identify as systems a great many things that were not previously called by that name. Such widely dissimilar things as galaxies, organisms, and ecologies are now seen as so many varieties of systems: astronomical systems, biological systems, ecological systems, and so on. At first sight, this may appear to collapse the distinction between them: it seems to be reductionism in a new guise. Instead of reducing things to a concourse of atoms, as Democritus did, we now reduce them to the concept of systems.

In fact, there is no such fallacy involved in systems thinking. To speak of systems *per se* is, of course, a simplification, but it is not a reductionist one. Whereas traditional reductionism sought to find the commonality underlying diversity in reference to a shared *substance*, such as material atoms, contemporary systems theory seeks to find common features in terms of shared aspects of *organization*. Reductionism is comparable to looking at a barn, a home, and an office building as so many structures erected of brick and concrete, disregarding their particular differences. The systems sciences look at them in terms of the organization of the materials which gives each structure its specific characteristic. They discover repeating patterns in the organization, such as floors, doors, and windows, and evaluate these as so many variations on a common theme. But they do not hold that you can reduce a barn, a home, and an office building to sameness by taking them apart to individual pieces of brick and concrete. Such reduction eliminates precisely that which is essential about each structure: the organization of the materials into variously functioning wholes.

How is simplification possible in regard to organization? Consider that every theory generalizes certain commonalities underlying individual differentiations. The commonalities it abstracts are the recurrent features of phenomena—the nonvarying aspects of it: the *invariances*. The question is, which of the recurrent aspects of phenomena are abstracted as the basic and essential invariances? Classical science abstracted substance and causal interactions between substantive particulars. Contemporary science concentrates on organization: not what a thing is *per se*, nor how one thing produces an effect on one other thing, but rather how sets of events are structured and how they function in relation to their "environment"—other sets of things, likewise structured in space and time. These are invariances of process related to real-world systems. We shall call them *invariances of organization*.

There are invariances of various degrees of generality associated with almost anything we encounter in experience. Take a human being, for example. On the narrowest level of generality he or she manifests the invariance of a certain family characteristic, inherited from parents or due to upbringing within the family. Somewhat more extensive invariances characterize his or her physiology and syndrome of personality traits. A person is easygoing or ambitious, loving or indifferent, lean or chubby, pale or sanguine—and so on in a multitude of respects. He or she is also a businessperson or a teacher, a nurse or a soldier. These are wider invariances, shared with increasingly large groups of people. In addition a person is a citizen or subject of a given country and, last but not least, a member of the human race. This is the ultimate invariance we can associate with the concept "human being."

But "human being" is not the ultimate concept we can apply to an individual. We can take another step and say that our subject, in addition to all of the above, is also a "living being." Now we encompass in our definition millions of species of animals and plants, on land and in water. Certainly, we couldn't very well identify ourselves, or any other individual, simply by calling him or her a "living being"—a generality this wide fails to distinguish a person even from a sea urchin. But it does grasp an organizational invariance: what the structure and function of a person has in common with that of sea urchins and all other forms of life. And that can be very important when we try to understand his or her origins and present nature and role in the scheme of things.

The concepts "life" and "matter"—or organic and inorganic—have lost much of their usefulness as the ultimate and defining categories of what things are in the light of organizational invariances. For one thing, there is nowhere to draw the line between the living and the nonliving. The famous amoeba is a single-celled "animal" satisfying the criteria of life: metabolism and reproduction. As all other things we call living, the amoeba ingests substances from its environment, disposes of its waste products, and even reproduces itself. Thus in basic respects it behaves like any of us. But a virus is not so easy to classify. When in contact with the body of a host organism, it behaves as a living thing. When removed from such a body, however, it takes on the characteristics of a complex crystal.

This is not the only reason why it is pointless to try to draw hard and fast lines between the living and the nonliving. A further reason is that many things that we used to call non-living (or inorganic) manifest organizational characteristics which they share with living things. They take in and put out substances or energies; they maintain themselves amidst changing circumstances; and some even grow and evolve into different and more complex shapes. A candle flame maintains its form amidst a flow of energies and substances, yet one would not call it living except in a poetic metaphor. So does a waterfall, a storm center, a city, an ecology, a university, a nation, and even the United Nations. Yet these, too, we would not willingly call living. Hence there may be further organizational invariances which relate some inorganic, all organic, and most *supra*organic or social things. We could characterize any of these things by a concept which denotes and defines this underlying invariance. Such a concept tells us even less than "living thing" does about the individual characteristics of any given person but it tells us more about her origin, nature, and role in a wider context. The proper term for this highest level of organizational invariance is "natural system." In this use "natural" contrasts with "artificial" and not with "social." Any system which does not owe its existence to conscious human planning and execution is a natural system—including humans themselves and many of the multiperson systems in which they participate.

The concept of natural system is vast and its content is correspondingly general. Yet it is not an empty concept, for we can say things about natural systems that set them apart from other things, which are not natural, and/or have no systemic characteristics. Thus "natural system" is not just high-sounding but meaningless mumbo-jumbo, but has real content. It does not apply to a chair, a watch, a piece of rock, or a house.

We call a human being a natural system. We likewise call atoms, molecules, cells, organs, families, communities, institutions, organizations, states, and nations natural systems. But are we not collapsing the distinctions among them? We are not, for we do not claim that "natural system" describes everything about these entities. It is literally impossible to describe everything about anything, but the concepts of lesser generality describe more about fewer kinds of things than do those of greater generality. We are merely setting forth the same kind of generalization that we perform when we speak of a person as a human being and as a living thing.

There are important aspects of each of us which are expressed in the concept human, and others of a more basic kind which are defined by living thing. In the same way, aspects of a still more fundamental kind are expressed in the concept natural system. These are the aspects which all phenomena of organized complexity in nature have in common, and which permit us to speak of them in reference to a general concept. A human is a natural system if he or she exhibits the characteristics which other natural systems exhibit. These characteristics tell us something very fundamental about that person. They define his or her "nature," to use a now somewhat dated terminology. At any rate, they tell us what kind of a thing that person basically is, by showing what her psychological, physiological, and social organization has in common with a vast group of natural phenomena.

The more general a concept, the more widespread the invariance which it grasps. It tells us less about the individual peculiarities of a thing and more about what it shares with other things. In systemic thinking if you want to know what is truly fundamental about a human being, you define him/her as a natural phenomenon of organized complexity—a natural system. If you then inquire what sets off a human from fellow natural systems, you specify the many criteria which apply, first to organic systems as such, and then to the human as one species of organism. Leaving the realm of general theory, you can further specify a given individual as such-and-such a variation on the psychological, physiological, and social characteristics of *homo sapiens* until you have her defined as a unique person. This is the method of definition by specification, and it shows that detail and generality are inversely related. More importantly, it shows that *detail is a specification of some more general trait and must be comprehended within the latter as the relevant context.*

We don't know the basic nature of our son until we know him as a system arising in nature with properties shared with all other such systems; his unique characteristics are but specifications of these properties. When he cries because a favorite toy has been accidentally switched to a lower shelf, you encounter traits of specific individuality. When he laughs because his older brother repeated the family joke, you meet with a family characteristic. When he comes home from school with knowledge and pride in the work of his class, you encounter a social characteristic. When his throat hurts because of inflamed tonsils you are confronted with a trait of human physiology. And so on, down the line until we come up against the basic structuring of energies, information, and substances which permits the child, as a system of organized complexity, to counteract the wear and tear on his components and grow rather than decay.

We are natural systems first, living things second, human beings third, members of a society and culture fourth, and particular individuals fifth—we can make our own classification along such lines. In any case, we know ourselves if we know how basic characteristics of organized nature are specified to issue in that *sui generis* individual which each one of us turns out to be on close acquaintance.

3

*The Systems View
of Nature*

If we grant the importance of invariances of organization in knowing that remarkable segment of nature which exhibits the traits of organized complexity, we can proceed to outline some fundamental features of natural systems as such. These are the organizational features which all holistic phenomena have in common, and which each specifies according to his and her own genus, species, and personal individuality.

Before suggesting some of these features, however, we have one more question to answer. It is this: how can we ever come to discover the common features of the organization of natural systems? Obviously, if we had to resort to the method of examining each and every one of the things we suspect of being natural systems and comparing their individual traits, we would have a task well beyond human endurance and intelligence—even when aided by computers.

Fortunately, we have an alternative method. This may be surprising to those who think of all theory formulation as based on the classification of relevant observations, but it is nevertheless the one actually used in the advanced contemporary sciences. It is the "hypothetico-deductive" method: that, namely, of setting up a hypothesis as a working tool and then tracking it down to see whether it works in experience. This means that instead of asking, "What are the common observed characteristics of all things we call natural systems?" we ask, "What are the characteristics any observed thing must have if it is to be considered a natural system?" We formulate the properties of natural systems in abstraction, and then proceed to find out if they are actually exemplified in some observed things. The great advantage of this method is its efficiency: we may not be right, but we know what we are looking for. If we don't find it, we can always modify the hypothesis. This is certainly preferable to attempting to catalog anything that we come across, especially since we are likely to come across a great many things.

What characteristics must any object have, then, if we are to consider it a natural system? The emergence of isomorphic theories and parallel concepts in various natural and social sciences over the last decades permits the formulation of the required organizational invariances. These invariances traverse the spheres of physical, biological, and social phenomena and apply to systems of organized complexity wherever they are found and whatever their origin. The propositions offered on these pages are a generalized summary of contemporary scientific findings, designed to bring into focus the picture which they offer us of nature as a whole.

We have four interrelated propositions, each grasping an organizational invariance. Jointly they specify key characteristics of social, biological, and physical entities in light of which each can be considered a natural system. That such a vision can now be offered without indulging in purely imaginative leaps is due to the remarkable parallelisms emerging in contemporary scientific theories. It testifies to the coherence of a new world view of encompassing range and integrated self-consistency.

Following the method just outlined, our strategy will be to state the four organizational invariances one by one, attempt to clarify their meaning, and then explore them in regard to the corresponding natural or social sciences.

Proposition One: Natural Systems Are Wholes with Irreducible Properties

"Wholes" and "heaps" are not mysterious metaphysical notions but clearly, even mathematically, definable states of complex entities. The decisive difference is that wholes are not the simple sum of their parts, and heaps are. Take, for example, a pile of rubbish. Adding another can or removing a pop bottle makes only a quantitative difference to the pile—it becomes that much bigger or smaller. No other characteristic of it changes. We can add the properties of the additional item or we can remove them, but the characteristics remain additive—in other words, they do not change the characteristics of the pile as a whole. The same applies in the case of a heap of bricks, a rain shower, or a casual crowd in a public place. One part more or less means adding or taking away that particular part's physical mass and manifest properties, nothing more.

But compare this informal aggregation with an entity having some formal structure built on the basis of an interdependence among its parts. The most basic such unit consists of two parts in communication, where the outcome is something more than the simple sum of the properties of each. Friendship and love are of this kind. Friends and lovers do not individually have all the properties of their relationship, for a relationship is not merely Harry's friendship for Mike and Mike's friendship for Harry, or John's love for Mary and Mary's love for John. There is also *our* friendship and *our* love which, as romantic literature never tires of telling us, is more than we are in ourselves. Consider Plato's great discovery: the dialectic. According to Plato, two people, by challenging and responding to each other, can come closer to the truth than either one could by himself. The outcome of such a dialectic is not merely the knowledge of the one added to the knowledge of the other. It is something which neither of them knew before, and which neither would have been capable of knowing by himself. Such a twosome constitutes a whole which has properties irreducible to those of each individual by him- or herself.

The contemporary examples are still more convincing. Psychologists study the character of small and large groups, *as groups*. Since people behave differently in small intimate groups than in large public ones, there are some things we can say about the behavior of people in groups that refer to the structure of the group rather than to the individuality of its members. The properties of the group are irreducible to the properties of its individual members (although not, of course, to the properties of its members plus their relations with each other). Computing the character of the group by computing the individual properties and relationships of each member would be both hopelessly complex and entirely futile. The group manifests characteristics in virtue of being a group of a certain sort, and may maintain these properties even if all its individual members are replaced. Hence one might as well deal with the group *qua* group. And this means dealing with it as with a whole endowed with irreducible properties.

Holistic characteristics appear to be widespread among observed entities. We have considered a few examples, such as heaps of bricks and groups of people, and now it is time to proceed more systematically. We can do so by following up our plan to explore each of the hypothesized characteristics of natural systems in reference to our experience of nature. But "nature" is a big word and we must clarify just what kind of phenomena we understand by it. The systems view does not recognize absolute categories into which various natural entities could be conveniently pigeonholed. Yet some working categories are needed to organize the evidence and demonstrate its specific relevance. Hence instead of the more usual categories—inorganic, organic, and social—we shall use *suborganic, organic,* and *supraorganic.* We shall mean by these "levels" rather than "categories" of reality, distinguished in reference to modes of organization rather than to essence or substance. In general, by suborganic we shall mean the subject matter of the physical sciences, by organic that of the life sciences, and by supraorganic the domain of the social sciences. By reviewing evidence pertinent to them, we can gain some assessment of the validity of the properties characterizing natural systems as a basic category.

(i) Are there entities in the *suborganic* world whose properties as wholes cannot be reduced to the properties of their separable parts? This is our first question, and some careful thinking is needed right at the outset.

Atoms were taken as the indivisible basic building blocks of physical reality until the advent of modern atomic theory, which showed that atoms are complex and divisible. Their elementary particles were next thought to be indivisible, but they, too, turned out to be capable of scattering into quanta of radiant energies corresponding to several subsidiary particles. In the search for the genuine rock bottom of material reality, the latest candidates are the most unmatter-like "quarks." They are not isolable, nor are they known to exist in other than composite states, in which they are thought to constitute the many nucleonic and electronic particles known to the contemporary physicist.

On the other hand, we do know that atoms exist as discrete structures. Each constituent of an atom has certain properties (some, such as *spin,* so abstract that only a mathematical definition can be given of it), and the atom as a whole has certain properties. And the properties of the atom are not reducible to the properties of all its parts added together. If we took the neutron, proton, and electron of a hydrogen atom and recombined them in any arbitrary way, chances are we would not get a hydrogen atom at all. The properties of the

27

latter equal the properties of its parts *plus* the exact relations of the parts within the structure. These are usually expressed in terms of fields of force potentials (such as electronic and nuclear fields). Microphysics would be a simpler science indeed if atoms were mere heaps, like piles of rubbish or streams of raindrops. But such is not the case.

(ii) The above conclusion applies also to organisms, the key entities in the realm of the *organic*. If we could, by some ingenious method, take apart an organism to its constituent cells, molecules, and atoms, and reconstitute it without killing it, we would only get back the same organism (that is, one with precisely the same characteristics) if we (or it—for some sponges are capable of it) coordinated every single cell with every other in just the same way as we found it. All organisms are made up of basically the same substances: cells composed of molecules, and molecules composed of atoms of carbon, hydrogen, oxygen, nitrogen, iodine, phosphorus, potassium, sulfur, calcium, sodium, chlorine, iron, and a few others. The difference between Caesar and the chimpanzee is not a difference in substance but in the relational structuring of the substance.

Even the brain, that most delicate and complex of all known organs, is not merely a lot of neurons added together. While a genius must have more of the gray matter than a sparrow, the idiot may have just as much as the genius. The difference between them must be explained in terms of how those substances are organized. Since the precise correlation of every neuron with every other is more complex than a human brain can comprehend (in fact, no system can process sufficient information to decode every aspect of another system of equal complexity—that is, nobody can fully and exhaustively know himself, or anybody as complex as himself), the brain as a whole, or at least its particular subsystems (hemispheres, areas, or lobes) must be treated as wholes having irreducible properties.

In the same sense as the atom, the organism, and the brain, the human *personality*—a vague but increasingly researched aspect of human beings—also constitutes an irreducible whole. Whatever a personality may be, as Jung, Goldstein, Maslow, and other leading psychologists point out, it is not the mere sum of our feelings, volitions, instincts, and conceptions. It constitutes an integrated unity of all these in mutual relation. Whether we admit that there is such a thing as a subconscious or not, we must admit that we do not possess, say, the capacity to love independently of the capacity to reason, to will, or to worry. All such traits of our personality interact and constitute an integrated "personality syndrome" which acts as a whole and has properties as a whole. This is what we call "my personality" or simply "me."

(iii) The holistic nature of *supraorganic* entities, such as groups, has already been discussed. Whether we take a group as a class of students taking a course in history, a group of politicians debating a question of policy, or a group of football players trying to pass the ball through the opposite team's defenses, we are dealing with coalescing properties of many individuals expressed in the particular properties of the whole. Here, too, what makes a group what it is, is not just its membership, but the mutual *relations* of the members. The fact that physical entities such as atoms provide communication between their parts in terms of the interaction of fields of force potentials, that things such as organisms provide parts-communication by physicochemical means, and that multiperson organizations establish communications of quite another kind, does not invalidate their holistic character. Thus the intercommunication of members in a business corporation can take many forms, from gestured signals and the spoken word to written, verbal, and mathematical symbols, perhaps transmitted over complex communications equipment, but in all its forms it remains *communication*, that is, effective, mutually qualifying interaction between the members. It is in virtue of such communication that social institutions and organizations can act as entities in their own right, and can have the characteristics which go with their unified mode of behavior.

We have taken a brief look at some key entities in the suborganic, organic, and supraorganic realm in reference to our first proposition: that natural systems are wholes with irreducible properties. We have found that things as varied as atoms, organisms, and multiperson organizations satisfy this criterion. We now proceed to outline and investigate the next in the sequence of the four basic propositions that define the organizational invariance of natural systems.

Proposition Two: Natural Systems Maintain Themselves in a Changing Environment

It is obvious that things do not remain as they are indefinitely. Wherever we look, things either develop and evolve, or run down and decay. There is very little that is constant which meets our eye; even mountains are subject to erosion and continents to slow drift. Most of the ordinary objects we know are incapable of counteracting the wear and tear on their parts: they run down or get disorganized with the passage of time. A factory is this kind of an object, and so is a car. Constant tending is needed to maintain them in their present state. Moreover, they need fuel fed from the outside to keep them going. Ordinary objects cannot obtain their own requisite energies to run and keep themselves in good repair. But natural systems can. How do they perform such a remarkable feat?

That the physical world, on the whole, is running down, is expressed by one of the most fundamental of all laws of nature, the so-called Second Law of Thermodynamics. It states that a quantity termed "entropy" can only increase in time in any isolated system. Now, entropy, or its negative, is a measure of the free energy available to a system in virtue of the way its components are organized. For example, a house with a full tank of heating oil and good supply of electricity is so organized that it has energy available to heat and light itself and operate a number of electric appliances. But the heating oil (as well as the electricity stored in the batteries) can be exhausted, and in time the house will grow cold and dark. Hence most houses are supplied with regular deliveries of fuel oil and with a continuous input of electricity from a power source. Then the process of running down is postponed, but not eliminated. For now the house needs to import its working energies from the outside, and it is a question of how long the outside supplies last. Fuel oils are fossil fuels which were generated a long time ago in the history of the Earth by processes which resulted in the accumulation of reservoirs of natural oil under the surface. These can, of course, be depleted. Electricity is generated by burning coal, which is likewise an exhaustible fossil fuel, or by some other force, such as a waterfall, being used to drive the generators.

The question is how long such energies remain available. Although some (such as nuclear energies) may be available for a very long time, no energies are given in limitless supply. Eventually all the free energies available on the surface of the Earth can be used up, and then the house—every house—becomes cold and dark with finality. (This presupposes that no more sunlight—a source of energy we import from space—is available, an event a long way off indeed.) The principle states that within any given isolated system free energy stored in virtue of the organization of the components gets used up and the system gets correspondingly disorganized. The house as an isolated system would run down rather quickly. The house coupled to the power supplies of a continent forms a system of a much vaster kind, with correspondingly longer life expectancy. And a house coupled with the Earth-Sun system is a very vast system, with tremendous reserves of energies. But all such systems run down eventually, however long it may take.

Ordinary objects run down unless they are fed energies and repairs or replacements from the outside. Entire physical systems cut off from other systems run down in this way, too. But there are exceptions to the rule, and these are found on the inside of the closed systems of which the Second Law speaks. The Law is permissive: it does not determine just *how* such a system runs down. It can do so very unevenly. In fact it is quite possible that it should run down on the whole, while in some areas or parts it should actually get wound up. That is, there can be subsidiary systems *within* the whole system, and these subsystems can get more organized as time goes on, rather than less. Of course, the rest of the system gets correspondingly depleted, and the sum of the energies used up is always negative—more energy is used up than is generated. The system as a whole gets disorganized, whereas some of its parts become increasingly organized at the expense of the rest. It is as though we used the electricity stored in the batteries of our house to produce more batteries. We concentrate our available energies in the new batteries, but use up more energy from the original ones to make them than we preserve. The sum of the electric power available to us has decreased, even if locally (in the new batteries) it has

31

increased. The whole house runs down but some parts of it wind up.

If any given thing is to maintain itself in proper running condition, it must act as a subsystem within the total system which defines its energy supplies. It must be so organized that it draws energies from its environment, and burns them up in running itself. That is, it must take in substances which contain energies in a form which it can use for its own purposes. It then puts out waste products in the form of used-up substances, impoverishing its environment to that extent. The energies gained can be used to run the subsystem—something which inevitably has to be paid for in terms of the total supply of energy—and to carry out the necessary maintenance work. All this is directly involved in sustaining a subsystem over an appreciable period of time. Natural systems must keep running just to stay in the same place.

The particular configuration of parts and relationships which is maintained in a self-maintaining and repairing system is called a "steady-state." It is a state in which energies are continually used to maintain the relationship of the parts and keep them from collapsing in decay. This is a dynamic state, not a dead and inert one. And it does not violate any of the principles of the physical world.

The technical definition of a natural system is "open system in a steady-state." *Openness* refers to the energy import activities of the system, which it needs to "stay in the same place," that is, to maintain its own dynamic steady-state. We human beings are open natural systems; so are the cells that compose our body, and the ecologies and societies which we constitute jointly with our fellow human beings and with other organisms. Hence we are effectively embedded in the world of natural systems. (This situation will be of significance when we discuss our future and our values.) Let us now review the characteristics of various suborganic, organic, and supraorganic entities and see whether they do in fact manifest the properties outlined here.

(i) The basic system in the physical realm is the atom. Atoms may be stable or unstable. If they are stable, their energies are so well integrated that they balance each other and the atomic structure maintains itself in space and time.

Unstable atoms have dynamic instabilities, usually due to a very complex structure, consisting of many protons in the nucleus and a corresponding high number of electrons in the shells.

Stable atoms are usually considered closed systems: they do not exchange energies with their environment, although they are affected by high energies and heat. Atoms of this kind effectively withstand the overall course of degradation of energy predicted by the Second Law. They arrest entropy within their own structure. The internal forces balancing the atomic structure are so vast relative to its size that few external forces can disrupt it. But heat is one of the forces that can penetrate the atom's boundaries and intense heat—as well as highly accelerated particles—constitute external forces which can undo stable atoms. Under such conditions the energies binding the atomic nucleus are exceeded by the energies introduced externally, and nuclear fission—and possibly fusion—occurs. In our nuclear devices such conditions have to be created artificially, whereas they obtain constantly in the interior of all shining stars, including the sun. One outcome of the process may be "nuclear transmutation": the conversion of one type of nuclear structure into another, with several nuclei of the former kind fusing as they create one of the newer. Energies which do not fit into the configuration of balanced energies in the new structure are released. They are the radiations which maintain a star's luminosity and account for its light and heat.

The atom's behavior under conditions of radiations intense enough to penetrate its boundaries but not so high as to smash its nucleus is quite remarkable. Under electron "bombardment" the atom absorbs the radiant energy and ejects a corresponding quantum of energy from its own structure, usually in the form of one of its electrons. It is said that the atom becomes "excited" when it absorbs the extra energy from outside, and that it radiates off the "excitation potential" when it readjusts to its normal (ground energy) state.

The processes sketched here are not usually considered in the light of open-system theory: they are short-lived events, interspersing vast stretches of uneventful existence in the life of stable atoms under conditions more temperate than those of stellar interiors. But they do show that atoms are able to maintain themselves in a changing environment. They keep themselves running entirely on their own unless bothered by excessive heat, or by high-velocity energy-carrying particles. And even then they can perform the adjustments necessary to keep going, either by a quick readjustment of their electronic structure, or by a complete nuclear reorganization of the entire set of atomic fields.

Thus, contrary to the overall tendency of physical nature, stable atoms do not run down. They maintain themselves and can even transmute into more organized ones. But of course even nuclear transmutations do not contradict the Second Law, since the sum of energy is degraded consistently with it, as excess energies are radiated off and become unavailable for further work. As a result stars burn themselves out, while at the same time their atomic populations, initially mostly hydrogen, become more complex and organized.

(ii) When we change the setting and look at organisms as we know them here on Earth, we find analogous processes of self-maintenance exhibited in much more explicit form. Organisms are open systems all through their existence. They could not exist for more than a few minutes without the constant intake and output of energies, substances, and information. Think only what the survival chances would be for any organism if all its intake and output channels were closed. No air, no water, no food, no sensory information, no disposal of wastes—in sum, no interaction or communication with the world outside. No organism could survive under such circumstances.

Organisms not only constantly take in and put out substances, energies, and information, even more remarkably, they undergo a slow but inevitable exchange of *all* their parts. Hence organisms are very much like candle flames or waterfalls, in that their input and output constantly replace and replenish all their parts. But, unlike candle flames and waterfalls, organisms are able to maintain their particular structure under a variety of circumstances. They can get their own fuel and make their own repairs even when conditions change around them.

Of course, drastic changes in the environment may be beyond the adaptive capacity of any organism. Humans can export their terrestrial environment to the surface of the moon, and thus compensate for that rather drastic change in living conditions, but they are irrevocably damaged by a brick falling on the head. Other organisms, such as soft bodied insects, are more resistant to such forces, yet less capable of avoiding them. But whether by skill or by constitution, all organisms maintain their own vital constancies within a given range of variation in their living conditions.

The most remarkable organic self-maintenance phenomenon is the process known as "homeostasis." The term, coined by physiologist Walter Cannon in 1939, refers to the precise regulative mechanisms of warm-blooded creatures. Their body temperature is maintained constant, notwithstanding variations in the surrounding medium, and so is blood pressure, sugar and iron concentration, and a host of other essential substances and conditions.

The highly developed organism regulates its own internal environment, much as a thermostat regulates the temperature of a house. For this it requires reliable information concerning conditions in its surroundings. This comes from sense receptors (eye, ear, nose, touch, and taste) which tell the organism all it needs to know about its vital medium. If conditions change perniciously, the organism can take steps to protect itself—remove itself if it can, or close its shell, or activate its defense mechanisms. The more delicate organisms require advance warning of threatening conditions and the skill to interpret the relevant sense signals. They must be able to predict to some extent what is likely to happen (as a rabbit can predict that he is likely to be attacked when he smells a fox), and see about taking preventive measures.

We humans, more than any other organism, have greatly refined such predictive and interpretive skills. In fact, we have come to rely on these skills to such an extent that many of our natural physical defenses have deteriorated. We can neither fight nor run well enough to survive under attack from major predators. But we have the skill to know what is involved in the attack, and can deal with it either preventively or aggressively through the use of tools and instruments.

Humans can now take care of all their survival needs by using their predictive and manipulative capacities.

The living organism keeps itself in running condition as long as it can, and performs repairs if it gets damaged: these are the processes of healing and regeneration. But very complex organisms are unable to keep this up indefinitely, and succumb to internal exhaustion even when relatively undamaged: the process of aging. To survive, such species have managed to develop a way to perpetuate themselves by a form of super-repair: reproduction. Instead of replacing a damaged or worn-out *part*, they replace the *whole*. This way the individual organism undergoes the familiar life-cycle of birth, maturation, and death, but in its course reproduces itself and thus keeps the species going. The individual now becomes like the ripple on the surface of a larger wave in the sea: the individual, like the ripple, is local and temporal, while the species, like the wave, is vast and ongoing. Yet all the ripples together define the curvature of the wave itself.

The state maintained in and by organisms is the steady-state. As we have noted, this is a dynamic balance of energies and substances, always poised for action. It is never a plain equilibrium, such as the state a watch reaches when it has run down. In fact, the organic steady-state more closely resembles a wound-up watch, with forces available to activate all needed processes. The remarkable feature of the organism is that, unlike a watch, it *keeps itself* wound up, and thus counteracts the general tendency of things to run down. It does so by taking in highly organized energies (water, air, nutrients, sunlight), and breaking them down, using the liberated energy to maintain itself and grow. It puts out a much degraded form of energy (used-up air, bodily wastes) which, fortunately, still contains energies in a form usable by some other species.

By drawing energy from the Sun and combining it with degraded animal substances, plants recycle the energies and make them usable again for more complex organisms. The whole of nature, ecologists point out, is something like a vast, self-regulating and recycling system, drawing energy from the sun and running itself without surpluses and with minimal waste. It is a beautifully balanced mechanism, comparable to a repeating series of terraced waterfalls and whirlpools in a stream, with the energy needed to bring the water up to the head of the flow supplied by the sun. The substances and energies make the rounds, time after time, being part of now this fall, now that vortex, being upgraded and degraded in turn.

(iii) The scene before our eyes shifts as we contemplate the supraorganic sphere, yet many of the essential elements reappear. Here entire organisms jointly constitute many shifting patterns, some more enduring than others. These patterns tend to form wholes with their own irreducible characteristics. They also exhibit a tendency to perpetuate themselves. This self-maintaining, self-repairing property as it exists in human groups is what interests us at present.

If people were to congregate into groups when and as long as they pleased, groups and communities would be ephemeral phenomena indeed. A football player could sit down, if so inclined, in the middle of a pass, and a soldier could throw away his weapon when he had enough of the war. But that is not how things are in reality. There are rules, regulations, and laws, and even principles by which we stand. Customs and just plain habit enter as factors, together with an innate tendency to conform to one's culture and society. Even informal gatherings obey some unformulated and sometimes consciously unrecognized rules, which keep them coherent over some period of time.

Although some groups have built-in obsolescence, for instance, one-shot training programs for executives or technicians to bring them up to date on some new development, most groups have some degree of permanence. They may have "thruput" (a stream of people passing through them), yet, their structure is conserved. For example, although a university course exchanges its student membership everytime it is scheduled, the course itself has some degree of permanence. There is continuity in the way it is presented and discussed. This, too, can change, not as a function of the change of students but rather of the professor's own insight. And since the latter changes at a slower rate than the former, there is some degree of pattern-maintenance in the way courses are given over the years.

There are various degrees of rigidity and flexibility in the organization of multiperson groups. The point is that there are some *conservative* elements associated with each. Even conspirators in a revolution swear to a code of honor and behavior—one that is quite different, of course, from that of the society they are attempting to overthrow. In a larger group, which produces its own livelihood and mounts its own defenses, there are quite a number of pragmatic factors which conserve the structure.

For example, in the economy there are norms, such as Pareto's "natural price" of goods and services, which impose a high degree of self-regulation on production, distribution, and consumption. Many economists speak of an equilibrium which the economy strives to maintain—a process which parallels rather closely the homeostatic self-regulation of the animal body. Prices rise with demand; high prices on goods make for greater profits and attract more people to produce the goods; hence eventually the supply equals or surpasses the demand and prices fall again. Then production is cut until, through a number of fluctuations, some kind of equilibrium between supply and demand is reached. Likewise with defense: threat from outside forces (or subversion from within) calls for the mobilization of armed troops. If the threat is overcome, the troops become an excessive burden on the state finances and are reduced again. Here the balance involves threat to collective security on the one hand, and defensive capacities on the other.

The political and juridical structure of a society tends to remain consistent with the need to regulate individual behavior in accordance with established concepts of justice and the objective demands of social existence. Adjustment of undue tension between overly strict or overly permissive laws and such concepts and demands occurs by means of juridical reform or, if radical elements gain the upper hand, through political revolution. National and international structures obey analogous constraints. They follow a definite course, conserve continuity within change, and impose norms of behavior on their members.

Like atoms under excitation and organisms under changing conditions, social structures adjust and adapt, maintaining themselves in a dynamic steady-state rather than in one of inert equilibrium. Like a truly self-winding watch, they have forces at their disposal to activate their various functions and remain in a state of readiness. Inert equilibrium is a sure sign of decay, in the supraorganic sphere no less than in the organic and the suborganic one.

Proposition 3: Natural Systems Create Themselves in Response to Self-Creativity in Other Systems

Self-creativity in the sense suggested here is not a mysterious quality, limited to entities with "spirit" or "soul." It is a response to changing conditions which cannot be offset by adjustments based on the existing structure. In this more modest sense, self-creativity is a precondition of evolution. If natural systems were merely to maintain the *status quo* throughout the range of circumstances they encounter there would be no evolution, no patterns of development, and nothing we could call progress. Things would either succeed in remaining what they are, or go under. The evidence indicates, however, that many things not only manage to offset the pernicious influence of changes in their environment but are capable of development. Natural systems evolve new structures and new functions; they create themselves in time.

Now, there are two forms of change and they must not be confused. One is a preprogrammed kind of change, such as the evolution and growth of the embryo within the mother's womb. All the information the embryo needs to grow is coded right into the structure of its genes. The embryo as such is not creative—it does not make up its own patterns of development but follows already established pathways. This kind of change is typical of the process called "ontogenesis," which is the growth and maturation of the young of self-reproducing species. The other kind of change is typical of "phylogenesis," meaning the evolution of the species, and not just their individual members, from one generation to the next. Phylogenesis is the creative advance of nature into novelty: it is the trailblazing self-transformation of entire species and populations of organisms. This is the kind of change involved in the self-creativity of natural systems. It signifies the ability of systems to generate the very information which codes their structure and behavior.

It may be well to pause at this point to consider a thorny problem connected with the concept of evolution. The problem is, does evolution have a purpose, fulfill a plan, or strive toward some definite end-product or final stage? Is there some general blueprint which all things strive to realize by nature, as the classical Greek philosophers were wont to assume? Or is it all a giant game of dice, with chance ruling the evolution of species without any deeper meaning or plan to it?

This question can now be answered with more confidence than at any previous time in the history of scientific thought. Instead of speculations and rather arbitrary *ad hoc* hypotheses to account for this phenomenon or that, we can now glimpse something like a logic of developmental pathways as such. This logic comes from the workshop of mathematicians, systems theorists, cyberneticians, and similar "specialized generalists." The general drift of their hypotheses is something like this.

Assume that there are a number of at least partially organized objects sharing a field, space, or surface—what topologists call a landscape. Each of these objects is open to some influences from the environment and responds to these influences. Hence each object influences all others, directly or indirectly, by communicating with its own environment. Now, as each natural system (for these are the "objects" we are interested in) receives and responds to inputs from its fellows, it provides new inputs for the others. And so each system constantly challenges the others by responding itself to such challenges. There is interdependence among the systems—as with points along a net, when one is displaced, all others suffer some displacement, corresponding to where they are relative to the moving point. In natural systems, of course, the points themselves move and produce movement affecting the others—active responses, not merely passive effects. In virtue of the connectedness of all points, there is coordination in the behavior of all systems, and an overall pattern sooner or later emerges.

Assume further that the individual systems strung along the net are capable not merely of repeating certain types of behavior, but of inventing new ones. We get a progressive modification of behaviors: one invention poses challenges as its effects reach other systems, and these respond by their own matching inventions. Since behavior is based on structure, there must be an evolution in the structures of the systems themselves. As a result, we find ourselves with the basic mechanism of evolution us we know it in the biological sector. There are inventions or "mutations" produced from time to time, and we can assume that they are produced rather randomly, that is, that chance determines which system produces what invention at which time. Yet they are being produced fairly regularly, and some inventions prove to be more compatible than others with the parallel (innovative) behavior of the systems with which they interact. All inventions are equal at birth, but some become definitely "more equal" than others later on. The result is that certain coordinated patterns emerge in the pathways of innovation among the systems. There will be successful and unsuccessful inventions, and the successful ones, like Broadway plays, will have long runs while the unsuccessful ones will close shortly after opening night. Continued development discloses the refinement of the successful innovations, the elimination and merging of the less successful ones. The result is an ongoing reduction of chaos and the patient emergence of discernible order all along the network of systems.

That the reduction of chaos is ongoing and the emergence of order patient does not mean that the process itself would be smooth and continuous. In nature, as in the human sphere, nothing works in an unbroken and strictly linear sequence. Processes build up until they reach critical thresholds; then they trigger sudden change. By contrast, incremental improvements are seldom of fundamental importance. They may adapt a system to is environment, but are not likely to change it in a radical and lasting way. The fact is that systems that evolved a complex structure have a great deal of instability, and they manage to persist in their environment by buffering out all forces that threaten to change their structure in a radical fashion.

There are change-buffering feedbacks within all open systems, be they simple biological or complex sociocultural ones. The homeostatic mechanisms of the human body are a particularly sophisticated example of such a feedback on the biological level: they correct and compensate for abnormal conditions in the external environment by corresponding changes in the so-called "milieu interieur," the body's own internal environment. The laws and police forces of human society are a clear-cut example on the societal level: they, too, correct and compensate for deviations from established norms by preventive or punitive measures.

Institutions of all kinds, from civil bureaucracies to the hierarchies of the Church, produce codes and regulations and devise punishments and disincentives to assure that established norms and orders are preserved. It would call for a great deal of insight and forward planning on the part of the leaders of such institutions to neutralize their self-stabilizing feedbacks and achieve change before there is a desperate need for it. Such insight (and the will to act on it) is mostly missing; even in this day and age the rule in both the private sector of business and the public sector of government is "crisis management" instead of anticipatory planning and preventive self-transformation.

Because of the effect of self-stabilizing feedbacks, the evolution of the net of interacting systems, though continuous, is full of stops and starts. It comes to a temporary halt when the open systems that populate the net are stable and well fitted into their environment. It becomes revolutionary when some systems are critically destabilized and either evolve or perish. The experience of a system within such a net has a striking resemblance to the life of a policeman: it consists of   long periods of uneventful boredom, interspersed by short periods of rapid-fire terror.

There is one more feature to add to this picture. We must allow that the sudden and occasional "fulgurations" that create fundamental (rather than incremental) change in the systems change not only their internal structure, but also their external relations.

Now, evolution in the net can bring the functions of neighboring systems so closely tuned to each other that they respond to changes in other systems as a team. They delegate different parts of the response among themselves, with collaborators specializing in carrying out specific tasks. We have difficulty in sorting them out as individual systems, since from the viewpoint of any outsider they behave as one: to any stranger they speak with a single tongue. Since what they do among themselves in carrying out their joint tasks gets to be rather complex, and rather beside the point as far as their joint effect is concerned, there will be good grounds for lumping them together and considering them as a single system. Hence for an observer with a human type of intelligence the patterns of development include the merging of formerly separate systems into more complex *suprasystems*.

Such suprasystem-formation need not be limited by anything other than the availability of participating systems. We can assume that existing supersystems may collaborate and form *super-supersystems*, and that these, too, form systems of their own, until the entire net adopts the character of a giant system. Of course, that ultimate system will have no input unless it, too, is part of a further net (in which case we cannot call it ultimate), and so the situation of the net *qua* system will be different from that of any system or supersystem within it. The systemicity of the entire net is the effect or outcome of the alliances of the systems within it. It signifies that most general state toward which all events within the net tend.

Where, then, is such a net going? It is progressing from a state of great multiplicity and little coordination to a state of highly coordinate general forms of order. The *many* become parts of a *few*, and the *few* form coherent connections which make them part of the ultimate *one* of the net itself. But the many do not cease to exist for all that. While they become parts of teams, and of teams of teams, they nevertheless retain some individuality of their own. They exist as definite subassemblies within the larger whole. Moreover their function need not be conceived purely mechanistically, like that of cogs within a machine; theirs could be a function more like that of a vice president in an organization. Such functions are not uniquely determined by the situation in which the individual finds himself; his ability to deal with that situation is likewise a decisive factor. Systems within other systems can have autonomy and freedom of decision in just this sense.

If this rather abstract scenario is a good analogue of evolution in the real world, we get meaningful answers to our question concerning the existence of a master plan in nature. If by such a plan one means something preestablished and realized by purposive manipulation, then the answer is that contemporary science does not know—and does not *want* to know—anything about it. But if by plan one means a recognizable pattern of development, then the answer is definitely *yes*. That things develop the way they do rather than in some entirely different ways is, within limits, perfectly logical and foreseeable. Among these foreseeable characteristics of development are increasing coordination of formerly relatively isolated entities, the emergence of more general patterns of order, the consolidation of individuals in superordinate organizations, and the progressive refinement of certain types of functions and responses.

In evolution there is a progression from multiplicity and chaos to oneness and order. There is also progressive development of complex multiple-component individuals, fewer in number but more accomplished in behavior than the previous entities. Evolution does go one way rather than another, and it keeps on going that way as long as it does not come into conflict with basic physical laws.

Nature is "permissive." It permits the local decrease of entropy (which means growth in structure and organization) if entropy (that is, disorganization) increases proportionately elsewhere. It permits the development of myriad forms and patterns of organization, and it selects from among those which happen to have come about. Evolution could have taken different specific forms—there is nothing necessary about a species called *homo sapiens*, for example. But the many forms it could have (and perhaps has taken) in the vast reaches of the cosmos cannot include forms contrary to the general trend of development. We cannot see how evolution could fail to push toward organization and integration, complexity and individuation, whatever forms it may choose for realization. Thus there is a plan, but it is not a preestablished one. It sets forth the guidelines and lets chance play the role of selector of alternative pathways for its realization. There is purpose without slavery, and freedom without anarchy.

We have spoken of systems as existing in the limbo of some possible world, and then assumed that this possible world is in fact ours. What is the evidence for making this assumption? As we did before, let us look to scientific theories of suborganic, organic, and supraorganic entities for elucidation.

(i) Atoms will again serve well as our prime example of suborganic systems. Do atoms create themselves and offer innovations as a means of meeting the challenges of their environment? To talk like this is certainly innovative, but not necessarily in a happy sense of the term. Before too many eyebrows are raised we should define just what self-creativity means in reference to atomic nature.

We have seen in the foregoing section that atoms can halt the increase of entropy within their own structure and can even reverse it in nuclear transmutation processes. This is permitted by the physical universe inasmuch as it involves an overall loss of structure and organized energy in the immediate surroundings, as in a stellar interior or a nuclear reactor. Suppose, now, that we take a population of atoms within such a sphere—and if it is to be a natural rather than an artificial sphere it would have to be a stellar interior—and assess the developmental processes in reference to the theory of self-evolving nets.

Astrophysical evidence indicates that there is a generally one-way buildup of elements during the chemical evolution of stars. Starting with the lightest of all elements, hydrogen, transmutation processes fuse the lighter into the progressively heavier nuclei, building up from hydrogen into helium, and into the still heavier elements. These processes take place within the space-time continuum within which stars condense under increasing gravitational pressure. Space-time is flat (Euclidean) in regions empty of matter. Where matter is found, however, the matrix becomes contorted in the four-dimensional patterns described in the geometries of Riemann and Lobachevsky. The entities called "matter" represent forces introduced into the space-time continuum. They include gravitation, electromagnetism, nuclear exchange forces, the forces described in the Pauli exclusion principle, and possibly others, as yet little understood. These forces interact within the space-time matrix and produce the phenomena of the observable universe.

Their interaction results in increasingly integrated levels of organization. It defines a definite arrow of time in the cosmos. The combination of gravitation, electromagnetism, and nuclear binding forces builds up atomic nuclei in space-time. Nuclei capture electrons by virtue of their fields of attraction and accommodate them in shells governed by the principle of exclusion. That principle states that only one electron can occupy a given position (described by four quantum numbers) around a nucleus. A second electron must have opposite spin or different orientation, or be excluded to a shell beyond. Since gravity brings about the increasing density of matter in stellar regions, and nuclear and electromagnetic forces combined with the exclusion principle build up integrated nonhomogeneous atomic structures, the very nature of reality conspires to build systems rather than distribute matter evenly, or condense it into featureless blobs.

Further gravitational pull leads to increased pressure and temperature. Nuclear fission boundaries are reached and surpassed. Now the existing structures fuse into more integrated units of greater complexity: the atoms of the heavy elements. These have a larger number of protons in the nucleus and electrons in the shells than the atoms of light elements. Their chemical valence likewise changes, and that means a change in their "social" behavior.

The process continues until atoms in the cosmos populate each of the classes of possible atomic structures, from hydrogen (atomic number 1) to uranium (atomic number 92). The more continuously a given atom participates in processes in dense regions of matter (that is, forms part of stellar masses), the more likely it is to evolve toward the heavy end of the scale. Although we find atoms of all varieties in the cosmos, there is a general irreversibility in the distribution: hydrogen gets used up and converted into heavier elements. Of course, the amount of hydrogen nuclei floating in interstellar space is large, and entire populations of atoms could have reached the final stages of their evolution before other populations could have even begun theirs.

Suborganic nature turns out to be rather different from the mechanistic universe of Newtonian physics. It is a dynamic realm of interacting forces, resulting in the emergence of systems of increasingly organized complexity. Physical nature is not a machine, just as organic nature is not infused with a separate life force. The patterns of development parallel one another, although they take place on different levels and exhibit different, in practice irreducible characteristics.

(ii) There is a definite pattern to the development of organic species as well, even if the agency of evolution here is mutation in a seemingly random sequence. We need only to consider the contrast between simple unicellular organisms and warm-blooded creatures such as ourselves to see that, if evolution is indeed going anywhere, it is going from the simple system to the complex one. Of course some species, such as the parasites, have lost complexity in their development, and there are still a large number of very simple organisms around, which evidently left the call for complexity unheeded. But we are not saying that there is such a call at all. Complexity of structure or function is not a *goal* of evolution; it is a *result* of it. There is no goal (or we know of none in the contemporary sciences), but there is a pattern all the same: the pattern of self-creating natural systems in interaction.

Biological evolution occurs on Earth, and where else in the universe we do not know. But that it would be a cosmic accident is hard to accept, not because of vestiges of ancient animistic and anthropomorphic beliefs, but because of our trust in the uniformity of nature. Like conditions give rise to like results throughout the cosmos: this is the basic credo of the contemporary natural sciences. For if they did not, that would be the end of physics, of astronomy, and of kindred sciences—the most cherished in our possession. It would be difficult, on the other hand, to prove that laws do *not* hold elsewhere, for one could always reason that there are laws that hold, only we have not found them yet. And humankind could probably never completely give up its belief in life elsewhere in the cosmos, since it could never explore all there is in the cosmos, nor indeed sample enough of it to draw general conclusions on the basis of observation. That we do not consider ourselves cosmic accidents, limited for some inscrutable reason to a small planet of a smallish solar system toward the edge of a galaxy, is not due to rampant mythologizing but to our belief that what has occurred in one place is bound to

47

manifest itself in another, provided the conditions are similar. And so we have good reasons to think that there is life elsewhere in the cosmos, even if we would be foolish to think that it is exactly like life on Earth.

We know that, given such conditions as have prevailed on our planet some five thousand million years ago, the already heterogeneous atomic populations formed many kinds of further associations among themselves. These were molecules of varying complexity. And by constantly providing inputs for one another, and responding to these inputs by self-transformations, new types of associations came about, some of which combined complexity with stability. These tended to be preserved. There need not have been predetermination of any kind as to precisely *which* structures formed and proved themselves capable of existence; it is enough that *some* did. Having formed, such systems changed the character of their surroundings by providing a fresh set of data for other proto-organisms around them. And thus the great game of mutual accommodation could begin in earnest, favored by energy conditions which permitted the bonding of atoms, such as carbon and oxygen, in long and complex chains. With such possibilities, the resulting structures could be fanciful indeed. And that they were is amply demonstrated by the subsequent evolution of life on Earth.

In the biosphere of the Earth, organic systems interact with one another, mutually eliciting creative responses. The progressive transformation of organic species pushes the front of evolution forward, exploring various forms and possibilities, each tending to be more complex than the foregoing. Some are successful, others fail. Even minor factors, such as a drop of a few degrees in the average annual temperature, can produce major effects, as modifications snowball and get magnified in the process. The demise of dinosaurs, after the longest undisputed reign of any species on earth, bears testimony on this point.

In biological evolution, organic systems join as partners within supersystems, and then combine with others in still higher-level systems. There is a progression from cellular, to to multicellular, to ecological systems . . . to the Gaia system as a whole.

Though this progression is real, we do not directly experience all phases of it: our sense organs are only equipped to distinguish systems on the multicellular level, for these are the systems with which we carry out our daily commerce of eating, drinking, resting, mating, and child rearing. That these systems are themselves composed of cellular systems and that they compose ecological suprasystems is not disclosed by our senses. These things we had to find out by reasoning, a fact which explains why we ignored them until recently. But we now see that the living things we know through our immediate experience are phases in the organization of the biosphere: they are *wholes* in one "cut" and *parts* in another. And their own parts are systems on their own level, and even *their* parts are that, until we scrape the bottom of the hierarchy with the atom and its doubtfully "elementary" particles.

(iii) Consider now evolution on the level of society. The ability of human social systems to survive depends in a very great measure on their ability to adapt to changing realities. Because they are culture-conditioned, social systems are embedded in an even more mercurial environment than biological systems. What the "reality" is that affects the existence of social institutions, states, economies, and so on, depends not only on what the case is, but on what its members, or its leadership, *believe* that it is. An institution such as the Catholic Church does not face a reality that is essentially different today from that it faced a thousand years ago—God presumably did not change, nor the verity of the Old and New Testaments. There are still people in need of salvation, and there are hardly any really new sins any more than there are new virtues. But how vastly different is the reality faced by the Church nevertheless! The difference lies in the minds of modern people, in their degrees of Christianity and strength of faith. If the Church, like any other social system, is to meet the challenge of this new reality, it must creatively transform itself to appeal as much to the modern mind as it did to the medieval one.

Likewise in the case of economic and political systems: the difference between a successful and a failing economy and government is due in large part to how people think about them. Attitudes, beliefs, worldviews—these all play a vital role

49

in determining the environment of social systems. Not that the real and objective factors can be neutralized, but they are overlaid by what people believe about them, and thus their effect is modified (cushioned or sharpened) by the dominant culture.

Because fashions in thought and belief, much like fashion in ladies' wear, are changeable, social systems are constantly challenged from within. Their own membership criticizes them and presses for changes. But this is only half the story. Social systems are also challenged from the outside, as other systems threaten to change or to dominate them. The most familiar of these threats is war. Armed forces are deployed to force the will of one politically organized system on another. Whatever the stated purposes of a war, the loser will not remain unchanged by its consequences. The same is true of the smaller-scale struggles between business corporations, social and educational institutions, and so on. While their very existence may not be constantly in the balance, social systems are always subject to pressures from within as well as from without, and must remain on their toes if they are to assure their own long-range survival.

Social systems form elements in an interdetermined net within their range of intercommunication. This range increases in time. Primitive tribal societies were relatively isolated, even from their own neighbors. They were self-sufficient entities, interacting mainly in the form of competition or occasional aggression. Even the more thoroughly communicating tribes did not exchange goods and women over large areas. They had neither the wish nor the means to travel long distances, and no technology to communicate over large spaces. Thus inhabited regions were relatively closed systems, as primitive civilizations evolved without major external influences. Surviving cultures in the central regions of Africa, Australia, and Borneo are the most clear-cut examples of closed regional systems today. More advanced civilizations had wider spheres of influence and vaster territories, as illustrated by the pre-Columbian Indian cultures of America. Still more prominent civilizations extended their range of influence through military as well as economic conquests. The Egyptians, Greeks, and Romans did so in classical times.

With the further advance of civilization communication capabilities between peoples improved, both by means of the physical transfer of persons on land and over sea and through the development of techniques for sending messages, first by courier, and later by technical means. Today the sphere of social intercommunication is all-embracing. There is hardly a tribe left that could exempt itself from it. The remotest regions of human habitation are readily surveyed from the air, readily communicated with by radio, television, telephone and through electronic networks, and can be made accessible for the exchange of goods.

There is another aspect of sociocultural evolution that merits attention. This is the agglomeration of smaller social units into bigger ones in a vast series of cooperative superimpositions. The smallest social unit for people is the nuclear family. Even in simple societies it is integrated within the larger structure of a clan or tribe. On this level each individual has manifold duties to perform and roles to play—father, hunter, advisor, and the like. Social development conduces, however, toward more complex forms of organization and with it toward the more clear-cut alignment of individuals within the social structure.

Social systems, like systems in nature, form "holarchies." These are multi-level flexibly coordinated structures that act as wholes despite their complexity. There are many levels, and yet there is integration.

While each individual belongs to a differentiated social institution, the social institution itself belongs to other, more embracing associations and systems. Small community administrations are part of larger regional ones, and the latter are components of still larger state or national governments. Small businesses have an increasingly difficult time making it on their own, and need affiliation with trade unions, boards, professional associations of many kinds, linking themselves to the economy of their country. The boundaries of a state or nation are only relative. Social and business institutions and corporations spawn overseas branches, and merge with, acquire, or become assimilated in foreign enterprises. Goods are designed in one country, manufactured in another, shipped to many more, and the profits administered in a third. A scare in one stock market causes prices to dive in others on the other side of the globe. The ambitions of one country have impact on all its allies and enemies, no matter how far away they may be. The world communicates practically instantaneously and becomes, in McLuhan's phrase, a global village. More exactly, it becomes a global holarchy.

If we take such a broad view of social history, what are the systemic characteristics that we find manifested? We find, first of all, the extreme versatility of most social systems. Under pressures from within and without, they change with the times. Those that do not are left to history as ossified relics of the past. Inputs from within or without call forth innovations, and the innovative system produces new kinds of inputs on all systems with which it communicates. Thus a change in one triggers changes in others. The changes tend to be in the direction of greater structuration and improved technology. They confer greater capabilities on the systems, including increased power to affect other systems. Further differentiation brings with it the agglomeration of units in vaster cooperative networks—the chain-store effect.

Formerly autonomous systems are now subordinated to control from above, without thereby fully surrendering their autonomy. Systems *align* themselves in suprasystems, rather than *disaggregate* into them. Smaller systems are not melted down and recast into larger systems; the small units are still there, and they exercise essential functions. But these functions are part of the order in larger systems, which in turn may belong to still more encompassing ones. The top-level system is often hidden from view at the bottom. In today's world, employees of a business, social, or political agency may not even know which top-level corporation, agency, or conglomerate owns or directs their outfit.

There is a discernible trend toward differentiation, growth, suprasystem formation, and complexity throughout the range of social systems. The pace of development accelerates as the rate and sphere of communication increases. In our day both processes approach limiting values, as communication is almost instantaneous to almost anywhere. The effect of change in one system, be it economic, social, political, cultural, or educational, upon all others is now greater than ever before.

There are strains and stresses in this world which traverse the globe and tax the adaptive capacities of the individual, creating what Toffler calls future shock. World government is still in the realm of dreams, but world organizations sprout forth in increasing numbers, and the United Nations furnishes at least a forum for discussion and communication among nations. International power blocs are forming, partially crisscrossed by international economic alliances and diplomatic relationships. There is increasing order on this "spaceship Earth," even if the welding of the new order is paid for by the financial collapse of the less adaptive organizations and the nervous collapse of inflexible individuals. When we compare today's societal scene with that of even a hundred years ago, we see the tremendous increase in interdependence, complexity, and differentiation. The basic laws of development hold true in the supraorganic realm, as they do in the organic and suborganic sectors.

Proposition Four: Natural Systems Are Coordinating Interfaces in Nature's Holarchy

Because the patterns of development in all realms of nature are analogous, evolution appears to drive toward the superposition of system upon system in a continuous multilevel structure traversing the regions of the suborganic, the organic, and the supraorganic. Organization in nature comes to resemble a holarchic pyramid, with many relatively simple systems at the bottom and a few complex systems at the top. Between them all natural systems take intermediate positions; they link the levels below and above them. They are wholes in regard to their parts, and parts with respect to higher-level wholes.

Individual systems within a complex system have the role of coordinating interfaces. They assume the liaison between those (lower-level) components of the system which they control, and those (higher-level) ones which exercise control over them. Their function is to pull together the behavior of their own parts, and to integrate this joint effort with the behavior of other components in the system. This is a function which all natural systems must perform if they are to maintain themselves.

(i) Holarchic structuration is evident in the suborganic realm, both in the buildup of the atoms of the elements and in their inclusion in stellar structures as parts. In any system involving several atoms, each individual atom fulfills the interface function of integrating its own subatomic particles and coordinating its joint forces with the similarly integrated forces of other atoms. In electronic bonding within a multi-atomic molecule, for example, the outer shells of the atoms are open to one another, as the electrons describe complex orbits (or set up complex wave patterns) around several nuclei. The individual atom becomes something of an abstraction; it is now an integral part of the total molecular configuration. Yet it is still distinguishable on its own, since the nuclear and electronic forces focus and interact in definite patterns which pinpoint a specific region within the molecular structure with characteristics of its own. For example, a water molecule is made up of one oxygen and two hydrogen atoms, and, although it represents a unity with characteristics which are not reducible to any of its component atoms, the latter remain functional subassemblies. The same holds true for even highly complex molecular and crystalline arrangements involving thousands or millions of individual atoms as components. Each atom integrates its own component forces and contributes their joint characteristics to make up the integral properties of the whole.

(ii) Within the organic realm there are systems on many levels, from the macromolecular to the multicellular. For the sake of clarity, we shall concentrate on two levels: those of cells and organisms. Cells integrate the many subcellular assemblies, such as the nucleus, the substances of the cytoplasm, Golgi-bodies, and the rest. Each of these subassemblies consists of molecular or crystalline structures, which in turn reveal an atomic constitution. Thus cells, whatever their nature, are alike in being complex integrated holarchies of atoms, molecules, crystals, and subcellular organizations of a more complex kind—no cell leads an entirely autonomous existence. Although some cells lead lives typical of independent organisms, even amoebae are integrated within the ecology of their particular medium, much like other organisms.

Cells which jointly form complex living systems show even more remarkable forms of interface function. On the one hand they continue to exist as integrated wholes of their own components, carrying out the already remarkably complex metabolic and reproductive functions of life, while on the other hand they fit themselves with high precision to the requirements of life within an integrated organic community. There are cells that transform themselves to make the tissues of muscles; these have different functional characteristics from, say, brain cells, or the cells of the cornea in the eye. Each variety of cell contributes the integrated energies and functions at its disposal to the community, and in turn receives an environment which enables it to survive. The cooperative endeavor of cells is so tightly knit in higher organisms that the community itself acts as one, and is in fact a system in its own right. The human being is one such system, and his or her sense organs are communities of lower level systems designed to bring him or her information about other multicellular systems in the environment.

Particular cells can provide, as can particular organs, fresh inputs to the systems they constitute. Thus they constitute an influence from within. They either succeed in reforming some strands of coordination in the whole system, or they cause their own destruction, whether by killing off the system that supplies them with their vital necessities, or by making it reject them. Medical textbooks are filled with cases where cells and multicellular tissues, organs, and other subassemblies fail to conform to the pattern of the whole organism, and either make for some degree of accommodation in the rest of the body, or come to a bad end. Uncooperative systems are either ejected from the body or, like cancer, eventually destroy it and thus sign their own death warrant.

(iii) There is holarchic integration on the supraorganic level as well. Contemporary population biology elucidates the many strands of essential ties which bind organisms to one another and to their integrated ecology. Most of the complex species, such as practically all higher-order vertebrates, live in some sort of social community. These regulate the behavior of the individual members by assigning relations of rank ("pecking order") which are followed in all the vital activities—feeding, mating, defense, and so on. But whether or not social organization within a population is pronounced, the existence of each individual organism is precisely regulated by the balanced relationships within the ecology of its region. The great cyclic processes of the biosphere can take place with the required degree of accuracy and refinement because individual organisms fulfill their interface functions with precision. They feed, mate, deposit wastes, and even leave their carcasses in the right place at the right time. They communicate with other organisms, both socially and ecologically, and integrate themselves within the balanced holarchy of the global Gaia system.

In human beings social (i.e., intraspecific) relations assume a particular importance. People can communicate more extensively and intensively with their fellows than can members of any other species: they invented language, a unique innovation in terrestrial nature. While other species communicate remarkably well, they depend on patterns of communication expressed in sound, gesture, and odor for carrying out their integrative functions. Humans alone devised the symbol, which permits them to overcome the confines of the here-and-now. We can communicate about the past as well as the future, about our immediate surroundings as well as about remote and even abstract and imaginary things. All these enter as communicational realities into the social systems we build and make these systems qualitatively different from the social systems of other species.

But such differences do not exempt us from the interface obligations of our holarchic situation. We continue to depend on the proper coordination of the integrated functions of the bodily components which make us an individual organism, with the functions demanded of us in our social capacities. Physiologically we are an individual whole, whereas sociologically we are an integrated (even if sometimes reluctant) part. And since we are endowed with consciousness, psychologically we are both whole and part—a duality which, when not recognized as an interface coordination, can lead to confusion and distress.

The holarchic duality of functions is evident throughout the multiple levels of human social systems. For example, the task of district managers or heads of departments in large corporations is to assure the efficient functioning of their department and to bring its results to bear on the functioning of the organization as a whole. This means exercising care and control in coordinating the activities of all persons under them, and in assuring that the department performs the assigned tasks efficiently. These tasks in turn have to be measured against those assigned to other departments in the corporation, and interdepartmental meetings, phone calls, memos, and other forms of communication are used to work out the proper correlations. Jointly the many departments, coordinated and integrated, constitute the corporation itself.

The corporation as such (through its appointed representatives) talks to other corporations and makes the deals which assure wider spheres of cooperation. Ordinarily employees do not talk to other *corporations*, any more than the organs of one person communicate with other *people*. Persons in charge talk to their own counterparts in other corporations, in their function of assuring the proper inputs and outputs of their departments. And so on, up and down the many levels of the corporate world.

Systemic units collaborate with units of their own level and form supraunits, and these, in collaboration with their own kind, form still higher-level units. Each unit performs efficiently, and is assured of its existence, as long as it links together the integrated functions of its own members with the analogous functions of its counterparts within the structure. The business entity that fails in either of these tasks slowly but surely phases itself out of existence.

Similar observations hold in the sphere of political organization. Lower-ranking representatives of the people face the interface responsibility of serving their electorates as well as the larger administrative body to which they have been elected. The higher-level representatives face the larger task of serving their own country as well as helping to maintain military, cultural, and economic balance in the world community of nations. These separate but interdependent interests are clearly reflected in the distinction between domestic and foreign policy. They are understood by enlightened leaders to be complementary faces of a general policy of internal/external balance which alone can assure the survival and development of their country.

The social sphere abounds with examples of multi-level organization with interface responsibilities attached to the management of the intermediate units. Platoons in an army, parishes in a Church, schools in an educational system, bear witness to the multi-level aspect of social organization. Interface functioning means not only preserving an efficient method of operation, but keeping it efficient under changed circumstances. Subsystem conditions may change, necessitating the reintegration of the parts in a modified structure. Suprasystem conditions may likewise change and call for the realignment of the integrated contribution of the system within the community of other systems. Domestic and foreign policy must remain flexible and innovative. And the dangers of becoming too fanciful are balanced by the dangers of becoming ossified. Success is measured by the system's ability to anticipate changes in its sub- and super-structures and to cope with them.

A holarchically (rather than hierarchically) integrated system is not a passive system, committed to the *status quo*. It is a dynamic and adaptive entity, reflecting in its own functioning the patterns of change over all levels of the system.

Nature, in the systems view, is a sphere of complex and delicate organization. Systems communicate with systems and jointly form suprasystems. Strands of order traverse the emerging holarchy and take increasingly definite shape. Common characteristics are manifest in different forms on each of the many levels, with properties ranged in a continuous but irreducible sequence from level to level.

The holistic vision of nature is one of harmony and dynamic balance. Progress is triggered from below without determination from above, and is thus both definite and open-ended. To be "with it" one must adapt, and that means moving along. There is freedom in choosing one's paths of progress, yet this freedom is bounded by the limits of compatibility with the dynamic structure of the whole in which one finds oneself.

4

The Systems View of Ourselves

THE philosopher-scientists of antiquity viewed the human phenomenon within a cosmic context and held that to understand humans one must understand their world. But following the rise of modern science, investigators tended to dissect general questions concerning human nature into specific problems to be handled by specialized research. The classical scientific method led to a vast number of highly accomplished theories concerning man's behavior, dispositions, and even his subconscious. But it also led to the fragmentation of our understanding of human beings. In the midst of all the complex special theories, we have gained little real insight into human nature itself. In fact, some theories would deny that there is any such thing, preferring instead to think of humans as a black box which correlates stimuli with responses. Opposed to atomism and behaviorism, the systems view links the human being again with the world (s)he lives in, for he or she is seen as emerging in that world and reflecting its general character.

In contemporary systemic thinking the human being is not a *sui generis* phenomenon that can be studied without regard to other things. He/she is a natural entity, and an inhabitant of several interrelated worlds. By origin (s)he is a biological organism. By work and play (s)he is a social role carrier. And by conscious personality (s)he is a Janus-faced link integrating and coordinating the biological and the social worlds. The human being is, in the final analysis, a coordinating interface system in the multilevel holarchy of nature. To know one we must know something about that remarkable slice of reality which, instead of running down, keeps winding up.

The human being is one module in the multilevel structure that arose on earth as a result of nature's penchant for building up in one place what it takes down in others. On multiple levels, each with its own variant of the general systems-characteristics which reflect the nature of the self-constructive segment of the world, systems interact with systems and collaboratively form suprasystems. The human individual is a part of a majestic cathedral of great complexity of detail, yet of sweeping simplicity and order in overall design. All parts express the character of the whole, yet all parts are not the same. This is the systems concept of nature, and it is a precondition of coming to know ourselves.

Let us explore this concept. We start at the beginning, for the human being did not enter on the scene by a special act of creation but has always been part of this universe. Not as a human, of course, but as phenomena which harbored the potentials for becoming human. And so we go back to the primeval *ylem* or prime matter out of which all things arose by gradual stages.

Our Cosmic Origins

Imagine a universe made up not of things in space and in time, but of patterned flows extending throughout its reaches. What flows is a mysterious, nonindividualized something we call energy. It flows along pathways structured by the metric of integral space-time. It flows smoothly, without crinks or wrinkles, over vast stretches of this cosmic matrix, and it becomes contorted in some regions. In these regions there are disturbances along the flows induced by the presence of electromagnetic forces. Some of the flows tie themselves into knots and twist into a relatively stable pattern. Now there is something there—something enduring—whereas before there was but a transitory flow. Here and there, energy forms recognizable patterns which endure in time and repeat in space. "Things" are emerging from the background of flows like knots tied on a fishing net. These are local actualizations of energies which remain put. The highly integrated focal points of such phenomena are the most basic kinds of energy patterns the human intellect can objectify against the background of space-time. They are the particles of matter; Einstein called them "electromagnetic disturbances" in the space-time matrix.

Let us suppose that there are a vast number of such knots tied across the reaches of space-time, and that these knots are at uneven distances from one another. They form not isolated units but parts of a continuum, and they communicate with one another through the continuum. Their primary mode of communication is attraction and repulsion, depending on the distance separating them from one another. Attraction is the dominant mode of communication at all but extremely close intervals, and thus the knots in relative proximity move closer together. Many of them come to be concentrated in such close quarters that ordinary attraction breaks down and more complex strains and stresses are created between them. Some of the elementary units achieve cohesion in balancing the energy flows that constitute them in a joint pattern. They constitute "super-knots" of a much more complex kind.

A population of such complex entities transforms the character of space-time in the region of their concentration. There arises a material object—a star. These macro-objects continue to be connected through the continuum on which they are superimposed, but now they act as integrated masses: they form complexes constituted by the balance of their joint attractions and repulsions. The relatively stable super-units thus emerging further associate among themselves. Eventually the entire universe is dotted with balanced knots-within-knots in space-time, affecting each other and reaching further orders of delicate balance. The universe itself takes on the character of a vast system of balanced energies, acting in some discernible form of cohesion. Thus the whole universe expands, or expands and then re-contracts, or maintains itself in a dynamic steady-state—we are not sure which, at this stage of theoretical cosmology.

In some cosmic regions—such as planetary surfaces—further processes of structuration occur. Neighboring nodules interact and accommodate one another's internal flow patterns. The new integration of already integrated energies results in more complex flows along relatively stable pathways. The pathways themselves are the result of previous integrations; they themselves consist of energy-flows of established pattern. But now they serve to channel fresh flows of energy and act as "structure" in relation to "function." Hence new waves of formative energy course over stabilized structures, produced by foregoing waves. And the process continues; the beat goes on. Established structures jointly constitute new pathways and these, becoming established as structures in time, serve as templates for the production of new systems of flows. The patterns become complex; the cosmic cathedral of systems grows.

The known entities of science are interfaces located on various levels of the rising cathedral. *Electrons* and *nucleons* are condensations of energies in space-time field, based on the integration of *quarks*. They in turn are capable of integration into balanced structures: stable *atoms*. Here the integration of diverse forces within the nucleus produces a positive energy which is matched by the summed negative energy of the electrons in the surrounding shells. Uncompleted shells make the atom chemically active, that is, capable of forming bonds with neighboring atoms. We thus get states produced by the integration of the energies of several atoms: *chemical molecules*.

The tremendous potentials of electronic bonding, as well as of weaker forces of association, permit the formation of complex *polymer molecules* and *crystals* under energetically favorable conditions. In some regions, under especially favorable conditions, the level of organization reaches that of enormously heavy organic substances, such as *protein molecules* and *nucleic acids*. Now the basic building blocks are given for the constitution of self-replicating units of still higher organizational level: *cells*. These systems maintain a constant flow of substances through their structures, imposing on it a steady-state with specific parameters. The inputs and outputs may achieve coordination with analogous units in the surrounding medium, and we are on our way toward *multicellular* phenomena. The resulting structures—*organisms*—are likewise steady-state patterns imposed on a continuous flow, this time of free energies, substances (rigidly integrated energies), as well as information (coded patterns of energies).

The input-output channels of organisms can further solidify into pathways of definite structure, and the nature of these pathways, plus the organic systems themselves, define the *supraorganic* (ecological or social) *community*. Ultimately the strands of communication straddle the space-time region within which the primary systems have come together, and those of its layers which provide conditions favorable to such structuration become organized as systems in their own right. We reach the level of the *global* (ecological, and on Earth also sociocultural) *Gaia system*.

Our Place in the Universe

If the above is a sketchy but basically valid account of development and relatedness in nature, it is far from flattering for our ego. Once, in our own eyes, the human being was the center of the universe, the glory of creation. All development tended toward the human form or, in a more static worldview, the human being was the highest expression of the genius of the Creator. With the advance of knowledge, humanity's central place in the universe was questioned. The Sun, regretfully, refused to honor our expectations of serving us by faithfully following its light-giving path around the horizon; instead, it made our own Earth its satellite. The solar system, too, turned out to be somewhat less than the vastest or most majestic creation in the universe—it was discovered to be a rather minor system on the periphery of a galaxy. Those who viewed with alarm the displacement of humans from the center of the universe found scant comfort in the finding that at least our galaxy proved to be a large one.

But the dethronement of man in our eyes continued. The development of species from common origins was recognized as a law of biological evolution, and the human person lost status as a categorically superior type of being. Whatever qualities humans possess must have been acquired in the course of their development from a lowlier status. And it became difficult to convince oneself that anything as *sui generis* as a soul could have been acquired in the process of evolution. Humanity had to accept being placed in the ranks of other species in the animal kingdom.

But our species possesses characteristics unmatched by other animals and these could at last be pointed to with pride, as attainments unique to him. Consciousness, abstract thinking, language, feeling, and the expression and embodiment of these in communicational realities such as written and spoken words, works of art and other objects expressing feelings and dispositions, and the many signs and symbols which serve to transfer meaning and guide behavior in the human world, are surely unique achievements. And as such, it was believed, they are superior to the achievements of other species.

While the aforementioned things and capacities are unmatched by other species on Earth, we must recognize that they are not unattainable in principle by some of them: chimpanzees have been known to delight in spreading multicolored paint on canvas, to make and use tools, and to learn a rudimentary sort of symbol language. But the ultimate blow to our anthropocentric pretensions is dealt by the realization that humanlike qualities are not necessarily "higher" achievements or signs of evolutionary progress. Evolution may not go in the human direction for other species, nor is that to be construed a failure on their part. It is more likely that the human pathway of development is one of those innumerable experiments which evolution tries, follows up if successful, and abandons if not. It is neither better nor worse than having long necks like giraffes, wings like birds, or a whiplike tongue like anteaters. Evolution may not "drive" toward humanoid qualities at all, even if it uses them under rather special circumstances. What evolution may be up to could be merely the continuing structuration of the biosphere through increased levels of communication between systems on one level, resulting in more integrated suprasystems on the next.

But why, then, did human consciousness, thought, feeling, art, and language come about in the first place?

Consciousness

The idea of consciousness raises untold havoc and heated discussion, often for no other reason than its being used in different senses. It can mean "subjectivity"—a subjective awareness of the experience of sensations. In this sense also our dog has consciousness. It feels pain, hunger, thirst, and the mating urge; it can be happy and sad and is generally endowed with an inner life. But in another sense consciousness can be more than subjectivity: it can mean the ability not only to have but also to be *aware of having* sensations. By simply stopping for a moment and reflecting on my own mental states, I can examine my own sensations: I not only perceive that red object on the desk but also *know* that I perceive it. And I can not merely feel sad but be aware of feeling sad, and so on with all (or most) of the sensations that make up my subjective mental life. I am not at all sure that my dog has consciousness in this second sense, intelligent as he is in other respects. Reflective consciousness, at last, seems to be a uniquely human property. And it is the basis for a long series of other properties, all of which presuppose in some way the ability not only to perceive and feel things, but to know that one perceives and feels them and hence to order them in the light of his purposes.

Now, it is quite impossible to explain subjectivity (consciousness in the basic sense) by reference to the particular structure and behavior of the human organism and its brain and nervous system. If we grant that people have subjectivity, we have to grant that chimpanzees and dogs have it, since they, too, are endowed with brains and organs for perception, and show signs of purposive behavior. But if we grant this, then we are forced to admit that all organisms possessing a nervous system and evidencing goal-oriented behavior have subjectivity.

In reviewing the simpler forms of life, we find sensitivity as well as purposive behavior without anything more than the rudiments of a nervous system, such as nerve knots or ganglia. Worms, for example, give every indication of feeling something when they come up against an obstacle or when they are squeezed, and they seem to do their best to escape from unpleasant situations and find more suitable ones. Why, then, should we say that they are deprived of subjectivity? And how about plants? Recent experiments show unsuspected sensitivities even in them. Certainly they perceive sunlight, temperature changes, obstacles, and so on, for plants react to all these. Moreover, they modify their behavior as a result: most plant species will grow around obstacles and respond to changes in the direction of sunlight by movement or by putting forth more leaves in sunny areas than elsewhere. Granted that plants do not have perceptions and feelings of the human variety, are their sensitivities to conditions in their environment not rudimentary instances of perception? And if so, do they not warrant the assumption that these perceptions are "felt" by the plant in some way that is at least analogous to, although infinitely less evolved than, our own sensations? In short, there does not seem to be any good place in the organic realm to draw the line between species endowed with subjectivity and species which are not.

It certainly stretches credibility to extend the notion of subjectivity beyond the level of multicellular organisms, but there is no viable alternative. If a free-living amoeba is granted to have some primitive sort of sensation which corresponds to the "tropisms" by which it orients itself in its ambient, does the specialized cell of the plant or animal organism lack them? There is nothing that the unicellular animal has in terms of structure and function that the cell of the multicellular animal could not match. Thus, however reluctantly, we must conclude that the cells which make up our body have levels of sensations of their own, corresponding, of course, to their own levels of sensitivity and not to the sensitivity of our entire organism (to *our* sensitivity).

Rather than exhaust the patience of the reader by posing the same question over again for each principal kind of natural system, let us consider only two far-fetched cases. Take first the atom. When it is bombarded by a particle, or wave of a frequency above its threshold of elasticity but below the energies needed to produce nuclear fission, it reacts by quantum jumps and the emission of an amount of energy equivalent to that which it absorbs. Is this not sensitivity of a definite kind? And should we say that the atom is merely an automaton and feels nothing in the process? Of course, it feels nothing like what humans feel. But it could well feel something like what atoms feel, a subjectivity of sorts, even if an undifferentiated one.

The other example concerns the supraorganic systems constituted by plants, animals, and human beings. When a swarm of bees is disturbed in its nest, it is the individual bees that are disturbed and it is they who respond accordingly. But is it not also true that the swarm is disturbed, as a rather vague but real entity in its own right? And is it not also the swarm that reacts by pursuing the intruder, and not just the individual bees? Of course, we don't have the individual bees *and* the swarm, any more than we have the individual cells of a body and the body. But when a dog snaps at a fly on his nose, even though all his body cells collaborate in the action, it is not just his body cells which snap, but the dog as a whole animal. And it is precisely in this sense that it is not just the individual bees but the swarm as a whole which sets out in pursuit of the unfortunate intruder that disturbed it.

The difference between a swarm of bees and a dog is one of degree, not of kind. The dog is a more integrated system than a swarm of bees, therefore it is more convenient in more respects to speak of the dog acting than his body cells doing so. Think how awkward it would be to describe a concert goer's reaction to Beethoven as the reaction of the cells in his nervous system, not to mention of the subcellular tissues and bodies constituting his nerve cells. In the same way it is more convenient to speak of a student body being riotous or bright or lazy than each individual student, and of a nation being upset rather than each of its citizens.

At given levels of integration the many achieve cohesion and speak with the voice of one. Since this is just as true for body cells and organisms as it is for organisms and societies, what right do we have to deny subjectivity to the latter when we grant it to the others? We must end by acknowledging that subjectivity is possessed by all natural systems whatsoever, although the grade of subjectivity differs from level to level and species to species.

This conclusion does not do violence to our common-sense mode of thinking; it only extends it beyond its habitual boundaries. Clear and specific sensations are still reserved for animals with evolved brains and nervous systems, for we know of the dependence of highly differentiated forms of awareness upon neural structures and functions. But there is no unique correlation between the nervous system and the capacity for subjective sensations. Having subjectivity does not depend on having a nervous system; and the absence of a nervous system does not mean the absence of subjectivity, only its downgrading into some global, more undifferentiated flux of sensations, perhaps vaguely resembling those of pleasure and pain. The capacity for such sensations is most likely a universal feature of systems arising in nature.

Although the universality of subjectivity throughout the realms of organized complexity is a conclusion flowing logically from the holistic philosophy of the contemporary systems sciences, rigorous experimental evidence cannot be marshalled either for or against it. In the final analysis we can only observe our own sensations. I am already guessing when I speak of those of my wife and sons. Yet if I don't believe that it is reasonable to assume that my own subjectivity is unique in all the world, I must deduce the subjectivity of others from analogies of their physiology and behavior. And in systems thinking these analogies stretch both beyond and below the human being, clear across the vast holarchy of natural systems.

When subjectivity is defined as the ability of a system to register internal and external forces affecting its existence in the form of sensations, however primitive they may be, we must conclude that subjectivity is universal in nature's realms of organized complexity. But this conclusion does not hold for reflective consciousness, the ability of a system to be *aware* of its subjectivity. Self-awareness, as contrasted with subjectivity, does not appear to be a universal property of natural systems. There are good reasons to correlate self-awareness with certain varieties of highly integrated nervous functions, performed only by the most evolved nervous systems.

There are relatively clear-cut behavioral indications that tell us whether or not an organism is aware of its own sensations. Organisms endowed with reflective consciousness are liberated from the world of concrete here-and-now experience and can enter a quasi-autonomous world of their own creation. Subjectivity is the slave of actuality: it registers actual events only when they take place. Sometimes we see an elephant but often we do not; there is little we can do about it on the level of actual sensations. However, there is much indeed that we can do about conjuring up things like elephants, and even more outlandish things like electromagnetic fields of force potentials and twelve-tone music, on the level of reflective consciousness. For now we move not on the level of actual sensations, but on that of a *monitoring* of sensations. By having a running representation of our sensations, and not only the originals themselves, we can come to know and classify many of them, and establish their interrelations. We can conjure up the copies at will and can even create ideal entities like numbers and other abstract concepts—pure "monitoring" phenomena without direct counterparts in the sphere of actual sensations. The conscious mind is able to develop language and abstract thinking and these are feats not within the reach of mere subjectivity.

It is relatively easy to tell whether any organism possesses reflective consciousness by noting whether it has developed a language and other symbolic modes of expression and communication, and whether it can transcend the limits of the here-and-now by making plans not directly triggered by actual stimuli. Man alone passes this test. Just because animals such as cats talk to each other—sometimes so vociferously that they disturb our sleep—and because they appear to plan their strategies in catching mice, does not mean that they have reflective consciousness. Their talk is sign communication in the form of actual stimuli, and not symbol communication on the level of abstract thought. Their strategies in anticipating which way a mouse will run are triggered by the sight and smell of mice and are not plans excogitated while sunning on a convenient windowsill. Without disparaging the intelligence of cats and offending cat-lovers, we can say that cats, the same as dogs and other animals, while having highly differentiated subjective sensations and precisely correlated responses, have not managed to evolve the monitoring capacity which represents the autonomous realm of reflective consciousness. They see, feel, and know, but they do not *know* that they see, feel, and know. Nor can they manipulate their seeing, feeling, and knowing by their own volition.

Are we not saying, then, that this remarkable capacity of monitoring and knowing one's own mental events is a truly spiritual phenomenon, something that, at last, sets the human being apart from the rest of nature? There is bad news on this score, too. While such uniqueness was suspected through the ages, in the last few decades we have come to realize that the information flows that underlie self-awareness are not supernatural or even very complex. They can be built into artificial systems, such as sophisticated computerized servomechanisms. All that is needed is a circuit designed specifically to read the signals of other circuits.

For example, a set of machines in a factory is geared to perform a certain function, say, painting automobiles. The proper functioning of the machines is registered in a series of electric impulses which are fed into a computer. The computer can be programmed to switch on a green light when the signals manifest a certain pattern and a red light when another pattern appears. Thus the computer monitors the functioning of the machines, and signals any discrepancy between the desired norms and the actual performance. Engine performance instruments on cars and planes do essentially the same thing. Computers can be designed to deal with nothing but such signals. They can have programs which make them produce specific signals whenever an irregularity of function is registered. These signals can be fed to activate repair mechanisms when the irregularity is registered. When all things are back to normal, the emergency procedures can be automatically shut off and the passive phase of monitoring can be resumed.

Other frills can be added to this procedure. The controlling computer can be endowed with the capacity to try various repair routines in succession until one of them works. It can be programmed to learn from this experience and the next time the irregularity occurs to go straight to the successful procedure. Moreover, some really sophisticated computers can be taught to devise their own routines improving on their programmed performance. Checker-playing computers, for example, can perfect their strategies until they consistently beat their designers. "Dynamic optimizing programming" is truly human-like in many respects. Yet it is not due to the presence of a soul or spirit in the machine, but to a separate circuit which doesn't "do" anything (such as manufacturing goods or propelling a vehicle) but is entirely confined to monitoring the performance of other systems and setting it right when needed.

It is evident that highly complex systems required to fulfill a variety of tasks with a high degree of precision require a monitor of this kind. Many artificial systems fulfill such functions in man-machine teamwork: a human operator monitors the machine's functions and corrects for eventual malfunctions. Some advanced machines do this on their own. But systems in nature cannot evolve to such levels of complexity and precision where a special self corrective monitor would be required unless they develop the monitor themselves. Only humans have succeeded in creating a sophisticated monitor—although higher primates have made a good beginning in this direction. Our accomplished monitor is the cerebral cortex:

the seat of self-reflective consciousness. Without the cortex we would be a reflectionless vegetable, well capable of having sensations and coordinating basic bodily functions, but unable to think about our sensations and to plan ahead.

Having reflective consciousness makes us unique among systems in terrestrial nature. But this uniqueness does not suggest a supernatural quality, only a combination of most improbable circumstances, which led the human species to rely on at first primitive, and perhaps accidental, monitoring abilities.

Consciousness does, indeed, confer selective advantages. A species of organism possessing it can plan actions, communicate the plan within groups, and carry it out in purposive teamwork. In developing the rudiments of consciousness, our ancestors exploded the limits of genetically programmed behavior. They learned to learn from experience. By reflecting on the events of a hunt, for example, they could abstract its relevant elements and compare them with other occasions. They could select the most successful pattern of behavior and adopt it. Mere subjectivity is bound to the immediacy of events; only consciousness can liberate one from his actual experience and enable him to control it by his own will.

On this selective advantage of consciousness we came to rely more and more. Our physical strength, instinctive skills and patterns declined, even the acuity of our sense organs diminished. There was no longer a direct need for them; the brunt of the responsibility for survival rested on abstract mental processes, that is, on intelligence. As Jean Piaget and other investigators pointed out, intelligence is the most efficient instrument of interaction between organism and environment when the interaction is far-flung and complex, and has to be accurate. Its scope is greatly extended in space and time: an intelligent being can think about things past and future, and things far away as well.

Culture

The abstracted components of the stream of human sensations are those which recur with some regularity. These are the invariances in the stream, and our ancestors grasped them first in terms of their sensory qualities. They reified the recurrent patterns of sensory experience, endowing with thing-hood clusters of sensations which conformed to a common pattern. Later they symbolized these invariances, representing them in sound and in picture. The rudiments of art were already developed at least 50,000 years ago among the Cro-Magnon, as cave paintings in Lascaux and elsewhere testify. Language itself may have developed gradually over a period of as much as 500,000 years. It evolved from expressive signs, such as animals use to communicate, to denotative symbols, typical of human languages. Whereas signs provide a stimulus which signals something of immediate significance in the communicator's environment, a symbol may have a meaning which is entirely divorced from the here-and-now. A ritual dance in some species of birds signifies the readiness to mate at this time and place, but a love song can speak of sexual intercourse and its attendant sensations in general terms, or refer to it in a distant place, far in the past.

Human language, in using denotative symbols rather than expressive signs, became an effective instrument for communicating meaning. It enabled our ancestors not only to survive, but to dominate their world. Existence became social existence, within the context of a common set of meanings, communicated by means of a common language. Culture was born, and elaborate forms of social organization created. We became a sociocultural animal.

In the light of such considerations it might be natural to draw the conclusion that culture is "nothing but" an instrument of human survival. Yet such conclusion, while natural, would also be hasty. Psychiatrist Victor Frankl told us that "nothing-butness" is the contemporary variant of nineteenth-century nihilism's "nothingness," and one is as fallacious as the other. Culture may have arisen in the battleground of the fight for human survival, but once it arose it took on a life of its own. Culture is more than a tool of human survival—it is a qualitatively higher phenomenon. Thinking rationally and feeling with clarity and intensity, coupled perhaps with faith and a conscious morality, is qualitatively different from behaviors to assure one's survival and the continuity of the species. Culture and survival functions must not be confused. The ability to survive is programmed into every existing natural system. Homo alone has developed an autonomous *culture*.

There is no evidence to support the claim that an evolved culture has biological survival value, nor for the different claim, that once biological survival is assured, the inevitable next step is culture. To hold that human culture is a goal inscribed on the very banners of biological evolution is without foundation. But culture did come about, and it did arise as a sequel to the evolution of a self-monitoring brain coupled with high-level sensitivities subserving biological ends. It seems that the self-monitoring capacities of the human nervous system, coupled with its sensitivity to the environment, emancipated us from the confines of sensory reality and placed us within a world we ourselves created. We could surround ourselves with ideas, modes of feeling, and beliefs which are only indirectly related to the experienced world around us. When these capacities first emerged they conferred selective advantage in an environment which included many physically stronger and faster creatures. But once the capacities were developed, we became utterly dependent on them.

It is a case of the sorcerer's apprentice. If one uses reason in tracking down one's prey and in defending the common territory, one cannot stop using it when gazing at the starlit sky; reason cannot turn itself off. It is likewise impossible to reserve one's mystical feelings and mythical beliefs for times when these serve some positive function—as in rituals which can take the place of real aggression—and become unfeeling and unbelieving in daily life. Once we started to use reason in some things, we became stuck with our rationality. And when we evolved the capacity to substitute imaginative satisfactions for real ones, we also became saddled with the capacity to feel, envisage, and to believe. It became as impossible to return to the state of nature as it was for Adam and Eve to return to Paradise—a myth which expresses this insight in metaphorical terms. The apple, not only of knowledge but of the many facets of culture, proved to have irreversible effects. As you cannot uncook a half-cooked egg, so you cannot unlearn a half-acquired truth. Since there was no turning back, we could only go ahead. Recorded history testifies that we did.

Humanity's phylogenetic development called for reflective consciousness as a means of species survival. But consciousness, when evolved, took over the direction of our evolution. The means became the end: the self-maintaining biological species was transformed into a cultural species sensitive to knowledge, beauty, faith, and morality.

Our evolutionary history determined that we become a cultural creature, but did not determine what *kind* of culture we would have. Hence our problem today is not whether to have a culture; it is what kind of a culture to have. And this requires some serious thought. The kind of culture we inherited from our fathers and grandfathers is beginning to challenge our ability to survive on this planet. If we do nothing more than blindly accept it, we may not be able to do what they did, namely, hand it down to our children and grandchildren: we may not have the grandchildren to hand it down to.

This raises the question, what is it that ultimately determines the nature of a culture?

Recorded history is a fragment of the social history of our species. During this fragment, as in times before, the vast organizational processes of differentiation, complexification and association of individuals within groups, and of groups within groups, has taken place incessantly. The pace has now quickened, now slackened. But it has never reversed more than locally, and never for long. On the whole, the same branching evolutionary development took place here as in the organic realm. But here it took place within the setting of cultures. And cultures can quicken the pace of development or slow it down.

There are many factors in a culture which accelerate or brake societal processes. Tool-using capacity is one such factor, one which in our culture developed into the vast resources of contemporary technology. Mores, customs, and laws regulating human interrelations and the exchange of goods are further factors. The speed and range of interpersonal communication is still another. But over and above these, there is one set of factors which exercises determining influence, for it is this set which influences the persistence, growth, or decay of any particular kind of technology, law, and communication. This is the set of *values* prevalent in a society. Cultures are, in the final analysis, value-guided systems. Insofar as they are independent of biological need fulfillment and the reproductive needs of the species, cultures satisfy not bodily needs, but values. Values define cultural man's need for rationality, meaningfulness in emotional experience, richness of imagination, and depth of faith. All cultures respond to such suprabiological values. But in what form they do so depends on the specific kind of values people happen to have.

In early cultures rational, emotive, imaginative, and mystical elements were interwoven in synchretic unity. Myth is part science, part art, part religion. How many millennia humankind lived with one foot on the solid ground of biological and physical reality and the other in the nebulous world of myth is a subject for speculation. But scientific thought in the West did not divest itself of myth until the beginning of the great Hellenic culture, some four thousand years ago. In a slow but seemingly inexorable process, the rational and the emotive-imaginative-mystical elements of myth were separated. One cohered into philosophy, first cosmological, then humanistic and social; the other into religion, literature, and art. The great split that led to the medieval distinction between moral and natural sciences, and later to the malaise of the "two cultures," was foreshadowed in the rivalry of Greek philosophers and dramatists. The global unity of previous cultures was gone, and never entirely recovered.

The golden age of Greek civilization was guided by the ideal of living the good life. It was succeeded by Christianism in the West, where the good life was shifted to the next: the kingdom of God. It was not until the beginnings of the modern age that the eternal order of things was again subjected to empirical and rational scrutiny and people began to adopt new values. First they proceeded on the assumption that there are a fixed number of goods, and these have to be distributed as equally as possible. The gain of one person is the loss of another. However, with the rise of modern science and new techniques of transforming energy in the production of goods, the ideology of fair distribution was replaced by the ideology of growth. The theory of early capitalism, as expressed by Adam Smith, was based on the realization that there can be economic cycles where one thing leads continuously to the next, and by the time the cycle returns to its starting point there has been a gain all around. These cycles—such as saving, investment, production, distribution, consumption, labor, and renewed saving—were thought to apply to material goods. The spiritual goods were to follow in their wake, when everyone had had enough of whatever he or she wanted.

Equipped with the technological applications of Newtonian science, modern capitalism led to an unparalleled growth in economic productivity. Its values were materialistically oriented: *the good* is a large production per capita, and *the better* a still larger production. But internal problems created by the economy's development prompted many thinkers to formulate alternative theories. Marx proposed the most radical one in declaring the need to abolish private property, the division of labor, and class differences. Others, like Keynes, suggested creating an economy with a sufficiency of goods within everyone's earning capacity. Their ideals still centered on material goods as the precondition of human fulfillment, whether the goods were to be obtained through corporate capitalism based on private ownership, or by a socialist economy based on collective ownership. The golden age was to come when basic needs had been fulfilled, and that meant more production and a more equal distribution of goods.

In the meantime Western culture reduced the death rate, did not immediately drop the birth rate, increased interpersonal communication, and transformed the face of the Earth in its image. This made for the consumption of natural resources at a compounded rate by a vast population. Whereas in centuries and even decades past, people thought that the sky was the limit, our generation is forced to realize that the limit is the Earth. There is just so much of it, and just so much of what is there is usable for human purposes. Not only can we not increase per capita production indefinitely, we cannot even duplicate its present rate in America and Western Europe in the rest of the world. All the people of the world cannot live as Westerners do today; the Earth is not rich enough for that. This is something entirely new for believers in the ideology of growth. Progress cannot lie in more and bigger; it must be redefined, and that means having a new system of values. But what is there on which to base our new values? This is the paramount question of our day.

The Nature of Value

It was fashionable, not very long ago, to profess thorough skepticism in regard to value norms, holding them to be nothing more than expressions of personal likes and dislikes. Similar to arguments concerning one's preferences for the mountains or the seashore, they were said to be emotive responses on the part of individuals and not facts capable of proof and disproof. Relativism in cultural anthropology, together with subjectivism and noncognitivism in philosophical ethics, pronounced a death sentence on normative values; scholars found no justification for them, and hence held that there is no point even in discussing them. The evidence for such skepticism was grounded partly in the observation that people hold very different values, and partly in the fact that there does not seem to be any reliable empirical evidence of value itself. One cannot see values, nor can one hear, touch, taste, or smell them. Hence it would seem that there are no grounds on which value judgments could be justified, other than personal preference. And while preference may be a good reason for you liking one thing and me liking another, it is no justification at all for our coming to agree which thing is best.

Skepticism concerning the objectivity of value norms may be rather too hasty, however. That many value judgments lack awareness of objective foundations may be true. It may even be the case that all value judgments in the history of our species have ignored objective foundations. This would still not preclude there being discoverable objective foundations for values and hence the possibility of arriving at informed and objective value judgments.

Objective value norms can be deduced directly from the contemporary systemic world picture. The basic notions can be briefly stated.

Values are goals which behavior strives to realize. Any activity which is oriented toward the accomplishment of some end is value-oriented activity. This includes such brutish things as sonar-guided underwater torpedoes, which are self-orienting toward contact with a large metal body such as the hull of a ship, and such superlative ones as the brush strokes of a great painter, which have as their end the realization of a masterpiece on canvas. Nothing that pursues an end is value-free. Even science, that oft-cited paradigm of human objectivity, turned out in the light of recent investigations to be value-oriented not merely in the general sense of pursuing truth, but also in the specific sense of pursuing certain selected avenues toward the grasp of truth. There is nothing in the sphere of culture which would exempt us from the realm of values—no facts floating around, ready to be grasped without valuations and expectations.

Even more importantly, there is nothing in all the realms of natural systems which would be value-free when looked at from the vantage point of the systems themselves. Natural systems exploit the permissiveness of the physical universe in gathering in themselves order and usable energy at the expense of disorder and entropy in their environment. While the environment of such systems runs down, they themselves remain in steady-states and can even grow and organize themselves. If you happen to be one such system—as we all are—you find yourself with very definite goals. You must keep yourself running against the odds of the physical decay of all things, and to do so you must perform the necessary repairs, including (if you are a very complex system) the ultimate one of replacing your entire system by reproducing it from one special part of it. These are values common to all natural systems, shared in some form or another by suborganic, organic, and supraorganic systems alike. No system is free to reject these values for very long, for any change or reversal would result, with a high degree of probability, in its own disorganization.

But *how* these values are manifested depends on the specific characteristics and holarchical level of each system. Whereas all natural systems share a common value basis, the form in which these values are manifested differs from level to level and species to species. And when a species of system is highly diversified among its members, allowing for great internal variability of such things as nervous functions and consequent behavior, further specifications come about within the membership of the species itself. This is most strikingly the case with human beings. We do have the value base of all natural systems, and we specify it to fit the human level. And we do have the value basis of all persons but we specify it to fit our own individual thoughts and purposes. There is relativity here, but not a bottomless kind: it is measured against objective natural standards.

Contemporary cultural anthropologists are specifying a number of fundamental universal values, shared by people everywhere. The same basic values of survival, mutual collaboration, the raising of children, the worship of transcendent entities, and avoidance of suffering, injustice, and pain, are manifested by all cultures, albeit often in radically different ways. The surface forms differ, but the depth structures are analogous. The human being pursues his ends as a biological, social, and cultural being, wherever he lives.

If we survey the conclusions that emerge from these findings, we find that our objective basic values are those which we share with all natural systems. Each of us "must" (in the sense that he or she cannot help but) commit himself to survival, creativity, and mutual adaptation within a society of his peers; the alternative to these is isolation and death. But there is no imperative attached to the cultural specification of these values. These we can choose according to our insights. Of course, we "must" (in the same sense as above) remain within the limits of general natural-systems values. Finding and respecting these limits is precisely the problem facing us today.

To believe that "the Good" equals virtue in accordance with the harmony of the soul was an early specification of basic natural values. To believe that it equals production, with more and bigger of everything, is a contemporary view. To hold that affluence is good if it is based on a capitalist economy, or to believe that it is good only if it is based on a socialist one, are further specifications of contemporary values. The history of Western civilization was made by the contest over the specification of such materialistic values. There was disagreement as to *which* mode of production is the truly good one, and not *whether* production defines the nature of human good. To be sure, higher ideals are not easy to pursue on an empty stomach. On the other hand, there is no convincing evidence that great ideals *are* pursued when our stomach is filled. Thus filling it, while necessary, should not constitute our ultimate value.

It is easy to criticize. We all know that the critic always seems to know more than the person he criticizes. It is remarkable that relatively obscure critics should outrank great statesmen and scientists in intellectual acumen and sensitivity. If we take their criticisms at face value, that is just what they do. The rub is that it is one thing to point out the weakness of anything and quite another to produce such a thing. Both are needed, but criticism without construction becomes self-defeating: the critic ends by boxing himself into a state of cautious immobility. We must not stop, therefore, at pointing to the fallacies of existing values: we must point out new and better values. While this undertaking is bound to have many weak points of its own, it nevertheless has to be done. If it is to survive, even a turtle has to stick out its neck from time to time.

The Question of Norms

We are asking about the indicated values for human beings. Let us be careful to distinguish between *descriptive* and *normative* values. Our cultures do have values—a whole hierarchy of them. When described, they give us a system of descriptive values. We do not have, in any meaningful sense, normative values apart from these. Normative values (or value norms) are things we discover by examining human characteristics and pointing to those values which could lead people to fulfillment. Hence normative values are not described but *postulated*; they are creations of the inquiring intellect.

This is not to say that normative values are arbitrary. They are present in the sphere of actual valuations as norms expressed in various ways and to different approximations. A good example is the ordinary thermostat of a heating system. If you set the controls to 20 degrees Celsius and the mechanism is properly functioning, the actual heat of the house will fluctuate around 20 degrees. The actual heat level is like the descriptively known values of a culture; in turn, the setting of the thermostat corresponds to its normative values. There is no such thing as a separate normative value among all the descriptive ones, just as there is no separate temperature in the house for the thermostat setting. There are only actual valuations, descriptively or introspectively known. But they are regulated by intrinsic norms which we can discover by patient inquiry.

What are intrinsic human norms? The Greeks had an answer: they said that the end of the good life is happiness. Happiness, Aristotle specified, is the fulfillment of that which is specifically human in us. Typically for the rationalist temper of Greek civilization, Aristotle identified *reason* as the element which sets man apart from beast, and the fulfillment of which makes him happy. We can accept the Aristotelian ideas without this particular identification. Self-fulfillment, as contemporary humanistic thinkers and psychologists acknowledge, is the end of human purposeful behavior. It is the actualization of potentials inherent in all of us. It is the pattern of what *can* be, traced in actuality. Individual fulfillment can be a human value. And it can be specified and analyzed in the systems perspective.

Individual fulfillment is not the development of one faculty of the mind, or one part of the soul, as the Greeks would have it. It is the actualization of any number and any combination of different potentials, according to the temperament and conscious or unconscious desires of the individual. What is fulfilling for one may be constraining for another; we are different enough to make one person's meat another's poison. But we are not entirely different and we can talk of the syndrome of *human* potentials, out of which the paths of fulfillment of individual persons are selected. Fulfillment means the realization of human potentials for existence as a biological and a sociocultural being. It means bodily, as well as mental health. It means *adaptation* to the environment as a biological organism constituting an irreducible whole of its parts, and as a sociocultural role carrier collaboratively constituting the multiperson systems in his or her society. Fulfillment also means *acting on* the environment, both the internal one of the organism and the external one of the society, and making it compatible with the expression of one's potentials. It calls for a dynamic process of integration and adjustment, creating conditions for the actualization of the full potential there is in each of us for living, for knowing, for feeling, and for reaching toward the ultimate horizons of reality.

Individual fulfillment is a concrete process, conditioned by concrete factors. It takes place in the framework of the human situation, specified as the set of conditions that defines the existential reality of the given person. It is impossible, however, to specify norms for every situation in theory. Such can only be the task of *applying* the theory, with due regard to the specifics of the situation. We can gain some understanding of the overall type of conditions which specify the human situation, and such understanding can enable us to apply general theories in particular cases. Let us step back then and look at the general determinants of the contemporary human situation from the systems viewpoint.

Determinism and Freedom

In the world of organized complexity the arrow of time does not determine which pathway is taken by individual systems, only in what direction their paths converge. The general irreversibilities of organization include the progressive differentiation of existing systems, the merging of smaller systems within large unities without loss of individuality, and the increased level of communication among systems on their own hierarchical level. A corollary of these processes is the increase in the level of order in the largest suprasystem. In the language of information theory, this means that the amount of "noise" is reduced and replaced by "signals." In a process that includes any stops and starts and occasional fallbacks, the system itself becomes less open to chance, more regular and lawlike. Randomness is on the wane, determination on the rise.

There is empirical evidence that such indeed is the pattern of development in the sociocultural sphere. Relatively isolated and simple clans and tribes are incorporated in larger, more complex communities with an increase in communication among the incorporated units. The larger communities enter into communication among themselves, and jointly constitute still more embracing societies—nations, states, and empires. In our day we are approaching the outer limits of international communication and system-building. Further development, being unable to proceed extensively, will take effect intensively. Increasing communication among a finite number of national and multinational systems can only result in greater mutual determination among them. As the ratio of noise to signal is reduced through wider channels of effective international communication, the world will become more and more like a single unit.

A village is a unit because everyone knows everyone else and takes a role *vis-à-vis* the rest. The world will be a true "global village" when similar conditions prevail. Of course, on that scale the "everyone" who knows everyone else will not be individuals, but heads of state and heads of corporations in their formal and professional roles. Individuals will be more and more deeply embedded in complex social structures. Their interaction will be mediated by the interaction of the many sociocultural unities which they and their primary groups jointly form.

If this is the case, woe to the individuals. For to the extent that determination is introduced into the systems above them, they become determined as cogs in a fixed machinery. The more you organize, the more you regiment. To create the fully organized society is to create the fully regimented human cog.

No—this is a fallacious assessment of the situation. Although it is often voiced it commits one capital error: it conceives of individuals and society as basically mechanistic systems. In the new holistic view, however, they are dynamic, not mechanistic, systems: their seeming determinateness is not due to the casual *determination* of interactions among their parts, but to their *statistical correlation*.

In simple mechanisms each part receives an input which gives rise to an output in one way and one way only. You press a lever here, and up pops a tab there. No matter how many parts have been involved in the chain of events— gears, springs, shafts, etc.—the effect is determinate because each part transmits it in a rigorously determined way. If one part refuses to cooperate, the chain is broken and the tab fails to pop up at the end.

Natural systems are not at all like this. There are correlations between their inputs and outputs between what you press and what pops up—but these are not deterministic ones. The components of natural systems form something like emocracies in which it is agreed that certain functions will be carried out, but in which it is left up to volunteers to fulfill them. It matters not in the least which particular component carries out a task. What particular function a given component performs is also decided by the kinds of functions performed by the others. There is a high degree of internal plasticity within any natural system. The system as a whole is determinate, but the relationship of the parts is not. This is not the mechanistic casual determinism of classical scientists, but the flexible, dynamic "macrodetermination" of contemporary systems biologists, psychologists, and social scientists.

When a macrodeterministic system becomes a component in a system of its peers, its cooperative relations with its fellow systems take on the character of plasticity typical of parts in larger systems. Hence there is freedom (i.e., a significant degree of indeterminacy) on the level of electrons in the atom, and macrodetermination on the level of the atom as an integrated structure. There is such freedom on the level of molecules in a gas, but there is macrodetermination on the level of the pressure, volume, and temperature of the gas itself. And there is a similar freedom on the level of cells within a body, although there is macrodetermination on the level of the body itself. If you cut your finger it will bleed and subsequently heal, but it is not prescribed by any law of your physiology which particular cells are to form the epidermal layer, only that a certain number of them is doing so in a certain time sequence.

On the level of contemporary multiperson systems, macrodetermination is even more striking. We see corporations, universities, social organizations, and political regimes take on determinate structure, and we see that this structure does not impose mechanistic determinacy on their members. Sociocultural systems have openings for certain kinds of roles, from presidents to shoeshine boys. Persons with adequate qualifications can fill the jobs, regardless of their unique individuality. Roles are not made for given individuals, but for *kinds* of individuals classed according to qualification. When the roles are filled, the particular personality of each new tenant is reflected in his interrelations with others, and it produces corresponding shifts within the organizational structure. There is flexibility within the system, as part adjusts to part.

It is due to such plasticity that complex systems remain viable under changing circumstances. Totally mechanistic systems have only two states: a functional one where all parts work in the rigorously determined manner, and a failing one where one or more parts have broken down. They lack the plasticity of natural systems, which act as dynamic, self repairing wholes in regard to any deficiency.

The inverse side of macrodetermination is *functional autonomy*. The functional autonomy of parts within a natural system adds up to the macrodetermination of the whole. Functional autonomy does not mean independence. A fully autonomous (independent) set of units would not constitute a system, only a heap. Systemicity is imposed as a set of rules binding the parts among themselves. But these rules do not constrain the parts to act in one way and one way only; they merely prescribe that certain types of functions are carried out in certain sequences. The parts have options; as long as a sufficient number of sufficiently qualified units carries out the prescribed tasks, the requirements of systemic determination are met.

We can now appreciate why even an increasingly determinate set of social systems is not contrary to individual fulfillment. Fulfillment is predicated upon the freedom to become what one is capable of being—that is, upon the functional autonomy of human beings in society. Such freedom is a real possibility, although at present it is nowhere fully realized. To be sure, today's societies are not entirely mechanistic, but some are more mechanistic than others. Some designate the first-born of the leader to become the next leader, and thus exercise a mechanistic preselection of the fulfillment pathways for that individual. There are others which force persons to fill specific roles regardless of personal choices, whether in business, marriage, or politics, and they, too, are mechanistic to a corresponding extent.

Consider a system of national defense based on picking male offspring born to certain families to man the ramparts from a given day and month to another. The necessary defenses may no doubt be secured in this manner, but at the cost of personal freedom on the individual level and full rigidity on the level of the society. Such a mechanistic system can be revised to draft male offspring born in a certain year until the quota is filled. Now an element of chance enters, since whether any given individual is drafted for military duty is not determined in advance—only that a certain number is drafted from among a certain age group. But the mechanistic rigidity of the system has only been made more random, not more flexible. For the system still imposes a given fate on some individuals regardless of their preferences, and it still obtains a fixed number of soldiers.

Full natural-system macrodetermination based on the functional autonomy of individuals can be achieved in regard to national defense when the draft system is abolished and replaced by an all-volunteer army. Of course, abolishing such a system before there are sufficient volunteers to form an adequate army means the impairment of the defense capacities of the society, hence a malfunction on the level of the whole. But when the society is so structured that it offers channels of self-fulfillment within the ranks of its defense forces for a sufficient proportion of its population, the quota will be spontaneously filled. Society can exhibit macrodetermination based on the functional autonomy of individual persons.

Here lies the crux of the problem of our times. We are faced with the following variables: increasing communication—hence determination—on the macrolevel of sociocultural systems, great differentiation among individual aptitudes and potentials, and the value of individual human fulfillment. Our humanistic goal is to enhance individual fulfillment in an increasingly deterministic multilevel society composed of greatly differentiated individuals. Fortunately, this is a feasible endeavor. Like all complex natural systems, human institutions and societies function best when they are spontaneous expressions of the freely chosen activities of their interrelated members. Such a society is the norm against which we must measure existing forms of social structure.

What is needed is a reorientation of our cultural values in reference to the norms of individual fulfillment in a flexible and dynamic yet fully functional social system. To set such a goal we must have access both to empirical data and to the theoretically discovered norms of society. Empirical data are like the readings on a thermometer. They tell us what the case is, but not whether it is desirable. Theoretically discovered norms are comparable to finding the setting of a thermostat. They tell us how closely the actual values approximate the norms intrinsic to the system. We need both the readings and the norms. For only if we know both *where we are* and *where we want to go* can we act purposively in seeing about getting there.

The Western world offers the values of affluence as the panacea for all social ills. As norms, these values are now superannuated. In their place we must propose positive, humanistic value norms. Humanistic norms are not arbitrary: they are encoded in every natural system. But they can be overlaid by diverse cultural value objectives and hence, in times of urgency, they need to be consciously rediscovered. If they are found and adopted, we will again exercise our powers of adaptive innovation in maintaining ourselves and our culture within the thresholds of compatibility with the dynamic and balanced multilevel holarchy that is the biosphere-cum-humanity: the Gaia system of planet Earth.

A Role for Religion

Science addresses reason and intellect. Humanity, however, is both a rational and spiritual species; the human being has an intellective as well as an affective faculty. Consequently if the norms of contemporary humanism are not only to be discovered but also effectively brought to bear on the thinking and behavior of contemporary people, the rational discoveries of science need to be complemented by affective, basically spiritual, insights. Here is where the time-honored tasks of religion, as "re-ligio"—the binding and integrating of people within meaningful communities—takes on a fresh aspect. Not the inculcation of particular items of belief and patterns of action, but the basic orientation of human beings in the world around them is in question. This orientation should shift from the atomistic perspective of the mechanistic worldview to the holistic perspective of the systemic one.

Religions would not need to sacrifice, or even compromise, their cherished tenets to make an unique contribution to this shift. They would only need to draw on their own humanism and ecumenism to encourage creative thinking in regard to the elaboration and extension of their traditional insights. There is, obviously, a significant humanistic and ecumenical component in every great religion. Judaism sees humans as God's partners in the ongoing work of creation and calls on the people of Israel to be "a light to the nations." At the heart of the Christian teaching is love for a universal God reflected in love for one's fellows and service to one' neighbor. Islam, too, has a universal and ecumenical aspect: Tawhid, the religious witness "there is no god but Allah," is an affirmation of unity as Allah means divine presence and revelation for all people. Hinduism perceives the essential oneness of mankind within the oneness of the universe, and Buddhism has as its central tenet the interrelatedness of all things in "dependent co-origination." In the Chinese spiritual traditions harmony is a supreme principle of nature and society: in Confucianism harmony applies to human relationships in ethical terms, while in Taoism it is an almost esthetic concept defining the relationship between man and nature. And the Baha'i faith, the newest of the world religions, sees the whole of mankind as an organic unit in process of evolution toward peace and unity—a condition that it proclaims both desirable and inevitable. The great religions could draw on such ecumenical and humanistic elements to nurture a creative elaboration of their fundamental doctrines, supporting and promoting the shift to the new holistic consciousness.

The key unifying concept could be the spiritual assessment of the universe's progressive self-creation. The vast sweep of system-building processes from Big Bang to the emergence of life, mind, and consciousness could be recognized, and indeed celebrated by the religions. The recognition of the evolutionary self-creation of humanity, and of the larger reality of the cosmos, need not be confined to the empirical sciences. The process is all-embracing, and has a spiritual in addition to a physical dimension. We bear, after all, within our own body the impress of every transformation through which the universe has ever passed. The elements out of which our bodies are composed were created in the fiery processes of stellar interiors and stellar supernova bursts. They passed through a phased of dispersion in interstellar space, to be brought together in the womb of the proto-stars of a new stellar generation. As elements on the surface of the planets born of these stars, they participated in the original emergence of life in the rich mixture of molecules and protobionts in primeval seas. They entered and left living bodies for billions of years, cycling through the rich web of structured connections that now make up the reality of the self-maintaining and self-evolving Gaia system of this planet.

Not only our bodies, also our minds are immanent to this process. The forces that brought forth the quarks and the photons in the early moments of the radiance-filled cosmos, that condensed galaxies and stars in expanding spacetime, and that created the complex molecules and systems on life-bearing planets—these forces inform our brain and thus infuse our mind. They could come to self-recognition in each thinking and feeling human being.

By recognizing and celebrating the world's evolutionary self-creation, religions could promote this process of recognition in each individual. In light of their own grasp of the self-creative process, religious communities could celebrate the original flaming forth that gave birth to the known universe; the sudden synthesis of the quarks and of the wide diversity of microparticles as well as of atoms and molecules throughout the expanding reaches of cosmic space. They could celebrate the emergence of the macromolecules and protocells, the precursers and harbingers of life, on the surface of this planet and on countless if as yet unknown planets in this and in myriad other galaxies. They could celebrate the evolution of the noosphere on Earth as the next, and especially significant, phase in the world's evolutionary self-creation. They could help us recognize that our journey as individuals within the Earth's bio-noosphere reflects the evolutionary journey of the cosmos from Big Bang to black hole and again to a successive Bang; that the self-creating universe is our larger self—our primary sacred community.

Religious renewal always came in the wake of civilizational crises. It was in the disastrous moments of the history of Israel that the prophets of Judea made their appearance; Christianity established itself in the chaos left by the moral weakening of the citizens of a declining Roman Empire. The Buddha appeared in a period of spiritual and social confusion in India; Mohammed proclaimed his mission in an epoch of disorder in Arabia; and Baha'ullah wrote in confinement imposed by a moribund Ottoman Empire. Today, at a time when humankind is in the throes of the greatest and deepest transformation it has ever known, there is an epochal need for a creative extension of the traditional fundaments of the great religions, to complete and complement the rational worldview that is already emerging within the new sciences. With an alliance between science and religion, the shift to a systemic and holistic worldview would be reinforced. Both through reason and through feeling, contemporary people could be brought into closer harmony with each other, and with their environment.

In these critical times an adequate knowledge of ourselves and the world is no long a topic for academic discussion but an issue of deep public concern. Fortunately, contemporary science gives us a body of rational insights that are helpful as guideposts in finding the way to a humanistic future. If the insights would find a supporting echo in the spiritual dimension that has always been the essential domain of religion, the way would not only be found, but also effectively entered upon.

It is time to take stock. The systems view of the world is non-anthropocentric, but it is not non-humanistic for all that. It allows us to understand that we are one species of system in a complex and embracing holarchy of nature, and at the same time it tells us that all systems have value and intrinsic worth. They are goal-oriented, self-maintaining, and self-creating expressions of nature's penchant for order and balance. The status of the human being is not lessened by admitting the amoeba as his kin, nor by recognizing that sociocultural systems are his suprasystems. Seeing ourselves as a connecting link in a complex natural holarchy cancels our anthropocentrism, but seeing the holarchy itself as an expression of self-ordering and self-creating nature bolsters our self-esteem and encourages our humanism.

We may not be the center of the universe and the *telos* of evolution, but we are concrete embodiments of cosmic processes in their particular terrestrial variation. And we did evolve a most remarkable property: self-reflection. In virtue of this we may be among the very few species of natural systems in the universe which are able not only to sense the world and respond to it, but to know their own sensations and come to reasoned conclusions about the nature of reality. To be a human being is to have the almost unique opportunity of getting to know oneself and the world in which one lives. It is surely shortsighted to disregard this opportunity and confine oneself solely to the business of living.

A failure to exploit our capability for reasoned knowledge has now become contrary to the business of living itself. For our species may not be capable of existing for long without the use of rational insights in guiding its own destiny. Our knowledge has made us increasingly autonomous in nature, and enabled us to create the world of culture. It has freed us from many of the bonds of biological existence and given us license to determine our own evolution. But the possibility of error is the price we pay for freedom. The cultures we can build for ourselves may be manifold, but they must remain compatible with the structured holarchy of nature. We can build culture beyond these limits only at our immediate peril. Any such error must be rectified by using the same capabilities which originally led to the error: our relative autonomy conferred by our reflective consciousness.

Here is where the holistic vision of the systems sciences becomes important. It locates us within the multiple structures of nature and enables us to make constructive use of our capacities. Immersed in the immense structures of Gaia, we are nevertheless masters of our destiny for we have enormous control capabilities. As we regulate the organs and cells of our own body, so we must learn to regulate the many strands of social and ecological relations around us. We know fairly clearly what constitutes organic health for our biological system; now we must likewise learn the norms of our manifold ecologic, economic, political, and cultural systems.

The supreme challenge of our age is to specify, *and learn to respect*, the objective norms of existence within the complex and delicately balanced holarchic order that is both in us and around us. There is no other way to make sure that we achieve a culture that is both viable and humanistic.

The emerging worldview of the new sciences is embracing and relevant. When properly articulated, it can give us both factual and normative knowledge. Exploring such knowledge and applying it in determining our future is an opportunity we cannot afford to miss. For if we do, another chapter of terrestrial evolution will come to an end, and its unique experiment with reflective consciousness will be written off as a failure.

Selected Writings on Evolution and Society

I. STANDARD WORKS
[in alphabetical order]

Abraham, Ralph and C. Shaw, *Dynamics: The Geometry of Behavior*. Santa Cruz: Aerial Press, 1984.

Ashby, W. Ross, *An Introduction to Cybernetics*. London: Chapman & Hall; New York: Barnes & Noble, 1956.

Beer, Stafford, *Platforms of Change*. New York: John Wiley & Sons, 1979.

Beishon, J., and G. Peters, *Systems Behavior*. New York: Open University Press, 1972.

Bertalanffy, Ludwig von, General System Theory: *Essays on its Foundation and Development*. [rev. ed.] New York: George Braziller, 1968.

Blauberg, I.V., V.N. Sadovsky, and E.G. Yudin, *Systems Theory: Philosophical and Methodological Problems*. Moscow: Progress Publishers, 1977.

Boulding, Kenneth E., *Ecodynamics, a New Theory of Societal Evolution*. Beverly Hills and London: Sage, 1978.

Bowler, T. Downing, *General Systems Thinking: Its Scope and Applicability*. New York: Elsevier North Holland, 1981.

Buckley, Walter, ed., *Modern Systems Research for the Behavioral Scientist.* Chicago: Aldine, 1968.

Cavallo, Roger E., ed., *Systems Research Movement: Characteristics, Accomplishments, and Current Developments.* Louisville, KY: Society for General Systems Research, 1979.

Checkland, Peter., *Systems Thinking, Systems Practice.* New York: John Wiley, 1981.

Chaisson, Eric J., Cosmic Dawn: *The Origin of Matter and Life.* Boston: Atlantic, Little, Brown, 1981.

Churchman, C. West, *The Systems Approach.* (rev. and updated) New York: Harper & Row, 1979.

Club of Rome, Council of, *The First Global Revolution.* (written by Bertrand Schneider and Alexander King) New York: Pantheon Books, 1991.

Corning, Peter A., *The Synergism Hypothesis, A Theory of Progressive Evolution.* New York: Mcgraw-Hill, 1983.

Csányi, Vilmos, *General Theory of Evolution.* Durham and London: Duke University Press, 1989

Davidson, Mark. *Uncommon Sense: The Life and Thought of Ludwig von Bertalanffy.* Foreword by R. Buckminster Fuller, Introduction by Kenneth E. Boulding. Los Angeles: J.P. Tarcher, 1983.

Demerath, N.J., and R.A. Peterson, eds., *System Change and Conflict.* New York: Free Press, 1967.

Eigen, Manfred, and P. Schuster, *The Hypercycle: A Principle of Natural Self-Organization.* New York: Springer, 1979.

Eldredge, Niles, *Time Frames.* New York: Simon and Schuster, 1985.

Eldredge, Niles and Stephen J Gould, Punctuated Equilibria: an Alt-native to Phylogenetic Gradualism, in Schopf, ed.: *Models in Paleobiology.* San Francisco: Freeman, Cooper, 1972.

Falk, Richard, Samual S. Kim, and Saul H. Mendlovitz, (eds.) *Toward a Just World Order.* Boulder, Colorado: Westview Press, 1982.

Foerster, Heinz von, and George W. Zopf, Jr , *Principles of Self-Organization.* Oxford and New York: Pergamon Press, 1962.

Fuller, Buckminster. *Operating Manual for Spaceship Earth.* Carbondale,Ill: Southern Illinois University Press, 1970.

Gharajedaghi, Jamshid, *Toward a Systems Theory of Organization.* Seaside, Calif.: Intersystems Publications, 1985.

Glansdorff, P. and I. Prigogine, *Thermodynamic Theory of Structure, Stability and Fluctuations.* New York: Wiley Interscience, 1971.

Gray, William and Nicolas Rizzo, (eds.) Unity Through Diversity (2 Vols). New York: Gordon and Breach, 1973.

Haken, Hermann, *Synergetics.* New York: Springer, 1978.

Haken, Hermann, (ed.) *Dynamics of Synergetic Systems.* New York: Springer, 1980.

Jantsch, Erich, *Design for Evolution.* New York: Braziller, 1975

Jantsch, Erich, *The Self-Organizing Universe.* Oxford: Pergamon Press, 1980.

Jantsch, Erich and Conrad H. Waddington, (eds.) *Evolution and Consciousness.* Reading, Mass: Addison Wesley, 1976.

Katchalsky, Aharon and P.F. Curran, *Nonequilibrium Thermodynamics in Biophysics.* Cambridge, Mass: MIT Press, 1965.

Katsenelinboigen, Aron. *Some New Trends in System Theory.* Seaside, Calif.: Intersystems Publications, 1984.

Klir, George J., (ed.) Trends in General Systems Theory. New York: Wiley-Interscience, 1972.

Koestler, Arthur, and J.R. Smythies, (eds.) *Beyond Reductionism: New Perspectives in the Life Sciences.* London and New York: Macmillan, 1969.

Margenau, Henry ed., *Integrative Principles of Modern Thought.* New York: Gordon and Breach, 1972.

Maturana, Humberto R, and Francisco Varela, *Autopoietic Systems.* Biological Computer Laboratory, University of Illinois, Urbana, Ill: 1975.

Nappelbaum, E.L., Yu A. Yaroshevskii, and D.G. Zaydin. *Systems Research: Methodological Problems.* USSR Academy of Sciences, Institute for Systems Studies. Oxford and New York: Pergamon Press, 1984.

Nicolis, G. and I.Prigogine, *Self-Organization in Non-Equilibrium Systems.* New York: Wiley Interscience, 1977.

Pattee, Howard, (ed.) *Hierarchy Theory: The Challenge of Complex Systems.* New York: Braziller, 1973.

Prigogine, Ilya and I. Stengers, *Order Out of Chaos* (La Nouvelle Alliance). New York: Bantam, 1984.

Rapoport, Anatol, *General System Theory: Essential Concepts and Applications.* Cambridge, Mass.: Abacus Press, 1986.

Salk, Jonas, *The Anatomy of Reality.* New York: Columbia University Press, 1984.

Salk, Jonas, *The Survival of the Wisest.* New York: Harper & Row, 1973.

Simon, Herbert A., *The Sciences of the Artificial.* Cambridge, Mass.: MIT Press, 1969.

The Science and Praxis of Complexity. Tokyo: The United Nations University, 1985.

Thom, René, *Structural Stability and Morphogenesis.* Reading, Mass: Benjamin, 1972.

Varela, Francisco J. Autonomy and Autopoiesis. In *Self-Organizing Systems: An Interdisciplinary Approach.* Gerhard Roth and Helmut Schwegler. Frankfurt: Campus Verlag, 1981.

Weiss, Paul A., et al. *Hierarchically Organized Systems in Theory and Practice.* New York: Hafner, 1971.

Whyte, L.L., and AG. Wilson, and D. Wilson, (eds.) *Hierarchical Structures.* New York: American Elsevier, 1969.

Wiener, Norbert, *The Human Use of Human Beings: Cybernetics and Society.* (2nd ed.) Garden City, N.Y.: Doubleday Anchor Books, 1954.

Zeeman, Christopher, *Catastrophe Theory.* Reading, Mass: Benjamin, 1977.

II. SELECTED BOOKS BY ERVIN LASZLO
[in order of date of publication]

Essential Society: An Ontological Reconstruction. The Hague: Martinus Nijhoff, 1963.

Individualism, Collectivism and Political Power: A Relational Analysis of Ideological Conflict. The Hague: Martinus Nijhoff, 1963. [also in Japanese]

Human Values and Natural Science. (edited with J. Wilbur) New York and London: Gordon & Breach, 1970.

Evolution and Revolution: Patterns of Development in Nature, Society, Culture and Man (edited with R. Gotesky) New York and London: Gordon & Breach, 1971.

Introduction to Systems Philosophy: Toward a New Paradigm of Contemporary Thought. New York and London: Gordon & Breach; Toronto: Fitzhenry & Whiteside, 1972. reprinted: Gordon & Breach, 1984; second edition: New York: Harper Torchbooks, 1973.

The Systems View of the World: The Natural Philosophy of the New Developments in the Sciences. New York: George Braziller, 1972; Toronto: Doubleday Canada, 1972; Oxford: Basil Blackwell, 1975. [also in Persian, Japanese, French, Chinese, Korean and Italian]

The Relevance of General System Theory. (edited) New York: George Braziller, 1972.

Emergent Man. (edited with J. Stulman) New York and London: Gordon & Breach, 1972.

A Strategy for the Future: The Systems Approach to World Order. New York: George Braziller, 1974. [also in Japanese and Korean]

The World System: Models, Norms, Applications. (edited) New York: George Braziller, 1974.

Goals for Mankind: A Report to the Club of Rome on the New Horizons of Global Community. New York: E.P. Dutton, 1977; Toronto & Vancouver: Clarke, Irwin, 1977; London: Hutchinson, 1977;

revised edition: New York: New American Library Signet Books, 1978. [also in Italian, Spanish, Finnish, Japanese, and Serbo-Croatian]

Goals in a Global Community, Vol. I: Studies on the Conceptual Foundations. (edited with J. Bierman) Oxford and New York: Pergamon Press, 1977. *Vol. II: The International Values and Goals Studies.* (edited with J. Bierman) Oxford and New York: Pergamon Press, 1977.

The Inner Limits of Mankind: Heretical Reflections on Contemporary Values, Culture and Politics. Oxford and New York: Pergamon Press, 1978; revised edition: London: Oneworld Publications, 1989. [also in German, French, Italian, Chinese, and Korean]

The Objectives of the New International Economic Order. (with R. Baker, E. Eisenberg, and V.K. Raman) New York: UNITAR and Pergamon Press, 1978; reprinted: 1979. [also in Spanish]

The Obstacles to the New International Economic Order. (with J. Lozoya, J. Estevez, A. Bhattacharya and V.K. Raman) UNITAR and Pergamon Press, 1979. [also in Spanish]

The Structure of the World Economy and Prospects for a New International Economic Order. (edited with J. Kurtzman) New York: UNITAR and Pergamon Press, 1980. [also in Spanish]

Disarmament: The Human Factor. (edited with D.F. Keys) Oxford and New York: Pergamon Press, 1981.

Systems Science and World Order: Selected Studies. Oxford and New York: Pergamon Press, 1984.

World Encyclopedia of Peace: Vols. I-IV. (edited with L. Pauling and J.Y. Yoo) Oxford: Pergamon Press, 1986.

Evolution: The Grand Synthesis. Boston and London: Shambhala New Science Library, 1987. [also in Italian, German, Chinese, Spanish, and French]

The New Evolutionary Paradigm. (edited) New York: Gordon & Breach, 1991.

The Age of Bifurcation: The Key to Understanding the Changing World. New York and London: Gordon & Breach, 1992. [also in German, Spanish, Chinese, French, and Italian]

The Evolution of Cognitive Maps: New Paradigms for the 21st Century. (edited with I. Masulli) New York: Gordon & Breach, 1992.

A Multicultural Planet: Diversity and Dialogue in Our Common Future. Report of an Independent Expert Group to UNESCO. (edited) Oxford: Oneworld, 1992. [also in French, German, and Italian]

The Choice: Evolution or Extinction. The Thinking Person's Guide to Global Problems. Los Angeles: Tarcher/Putnam, 1994. [also in German, Chinese, and Hungarian]

Index

CPSIA information can be obtained
at www.ICGtesting.com
Printed in the USA
FFOW02n0654230115
10475FF

9 781572 730533